Mechanics of Granular Matter

Mechanics of Granular Matter

Qicheng Sun & Guangqian Wang

Tsinghua University, Beijing, China

Science Press

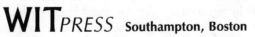

WITPRESS Southampton, Boston

Qicheng Sun & Guangqian Wang

Tsinghua University, Beijing, China

Published by

WIT Press

Ashurst Lodge, Ashurst, Southampton, SO40 7AA, UK
Tel: 44 (0) 238 029 3223; Fax: 44 (0) 238 029 2853
E-Mail: witpress@witpress.com
http://www.witpress.com

For USA, Canada and Mexico

WIT Press

25 Bridge Street, Billerica, MA 01821, USA
Tel: 978 667 5841; Fax: 978 667 7582
E-Mail: infousa@witpress.com
http://www.witpress.com

British Library Cataloguing-in-Publication Data

A Catalogue record for this book is available
from the British Library

ISBN: 978-1-84564-644-8
eISBN: 978-1-84564-645-5

Library of Congress Catalog Card Number: 2011936267

Contents

Preface

Granular materials are intrinsically athermal since their dynamics always occur at a state far from equilibrium. Quasi-static granular solids and granular flows are of great engineering importance, and innumerable equations have been presented to fit test data. However, their mechanical behaviours are still rather poorly understood, in contrast with the theoretical successes in studying highly excited granular gases.

Granular systems exhibit distinct characteristics on multiple spatial and temporal scales. A constituent particle is of course a solid, but granular materials may behave differently from ordinary solids, liquids and gases. This book focuses on the basic mechanics and underlying physics of granular materials. It starts with an introduction of contact mechanics of individual particles. It then discusses the structure of force chains network, and the influence on bulk mechanical properties of granular solids and granular flows. A preliminary multiscale framework is proposed for the nonlinear mechanics and strain localization in granular materials.

Special thanks are due to the following for their assistance: Prof. Jinghai Li encouraged Q.S. to conduct a multiscale analysis on granular materials. Prof. Feng Jin supported Q.S. during his difficult times. Dr Guohua Zhang revised the statistical mechanics. Dr Zhongwei Bi simulated the shear band initiation and development. Dr Xia Li and Dr Jianmin Qin provided the new derivations of stress and strain. Dr Shunying Ji and Dr Gongdan Zhou discussed the granular flow studies, and Mr. Jianguo Liu prepared photoelastic tests. The support of the National Key Basic Research Program of China, the Natural Science Foundation of China, and the State Key Laboratory of Hydroscience and Engineering, Tsinghua University, is also acknowledged.

Qicheng Sun & Guangqian Wang
Tsinghua University, 2013

Chapter 1

Behaviors of granular materials

Granular materials are large conglomerations of discrete macroscopic particles. The lower size (diameter) limit for particles is about 1 μm. The upper size limit may be a few meters, such as boulders in debris flows. In most circumstances, the forces between particles are essentially repulsive when there is a contact established between particles, and the effects of interstitial fluid can be neglected. One constituent particle is of course a solid, which can be described with classical mechanics, but granular materials may behave differently from ordinary solids, liquids, and gases. This has led many to characterize granular materials as a new form of matter. Granular mechanics have not only been important in engineering and geohazards for a long time, but also in the physics community in the past two decades.

1.1 Introduction

Many modern industries rely on the transportation and storage of granular materials. However, the technology for handling granular materials is not well developed, in comparison with the processing of fluids. An estimated 10% of energy is wasted each year in industrial productions. Hence, even a minor improvement in understanding how granular materials behave should have a profound impact for industries. In addition, the number of natural disasters, such as landslides, avalanches, and debris flows, has greatly increased in recent years. Figure 1.1 shows a large debris flow occurred in southwestern China on June 28, 2010. A geological survey in the summer of 2010 found that the flow generally comprised highly fragmented rock particles with a broad size distribution, from centimeters to meters. The debris flow is estimated to have run for 1,500 m at a high speed of 20 m/s, and covered the land to a depth of 10 to 20 m. A few key questions remain open, such as the mechanisms of the failure of highly fragmented rocks and the temporary-spatial scale of such debris flows. Obviously, strengthening the investigations on granular materials would be of paramount importance to production and the prevention and mitigation of such debris flows, but this imposes a huge scientific and technological challenge.

The study on granular materials dates back to Coulomb who proposed the ideas of static friction in 1773. Owing to the great engineering importance, innumerable continuum equations have been presented to deal with these materials. The existing mechanics of

Figure 1.1: Photo of the *Guan Ling* debris flow that occurred in southwestern China on June 28, 2010.

granular materials have been established on the basis of the representative volume element (RVE) stress–strain constitutive relation. From the engineering application point of view, it is necessary to revise the simplified rheological relations by introducing some phenomenological parameters, which are easy for numerical calculation, but are not suitable for explaining the physics of granular mechanics.

There are many notable names, such as Faraday who discovered the convective instability in a vibrating container filled with powder and Reynolds who introduced the notion of dilatancy, which implies that a compacted granular material must expand in order for it to undergo any shear. Widespread interest in granular media was aroused among physicists a decade ago, stimulated in large part by review articles revealing the intriguing fact that something as familiar as sand was still rather poorly understood. Sand has become a fruitful metaphor for describing many other systems, often including more microscopic and dissipative dynamical systems. De Gennes originally used sandpile avalanches as a macroscopic picture for the motion of flux lines in a type-II superconductor. A recent, intriguing use of this metaphor is based on the idea of self-organized criticality, originally described in terms of the avalanches in a sandpile close to its angle of repose. The resultant collective efforts since have greatly enhanced our understanding of granular materials, though the majority of theoretical considerations are focused on the limit of the highly excited gaseous state. Currently, granular matter has become one of the vibrant fields of study in physics and mechanics.

Figure 1.2 shows force chains within a granular assembly as observed in a static force penetration experiment. The existence of force chains is an important concept that has recently been put forward. It is a relatively stable quasi-linear structure subsequently connected by a few particles. In early photoelastic tests, Dantu [13] and Wakabayashi [2]

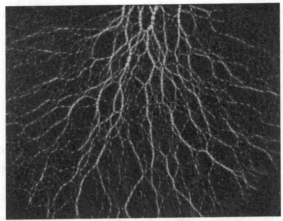

Figure 1.2: Photoelastic disks in a penetration test. (Force chains can be clearly observed.)

observed a tree-like force distribution. Bouchaud et al. [12] clearly put forward the concept of force chains "... arches are chain-like configurations of particles ... which act to transport force along the chains...." As shown in Figure 1.2, the stronger the contact force, the brighter the force chain.

It can be demonstrated that the normal components of contact force provide the major contribution to the deviatoric stress and that the spatial distribution of contact forces can be divided into two subnetworks: (1) contacts carrying forces less than the average (forming weak force chains) and (2) contacts carrying forces greater than the average (forming strong force chains). The contribution of the strong force chains to the deviatoric stress is dominant. These strong forces, carrying a greater than average normal contact force, are preferentially aligned in the major principal stress direction. Contacts that slide are predominantly within the weak force chains and they contribute primarily to the mean stress, with negligible contributions to the deviatoric stress. The weak force chains play a role similar to a supporting matrix by surrounding a solid backbone of the strong force chains.

Common materials are usually described as solid, liquid, or gases, but granular systems do not seem to exactly fall into any of these categories, and therefore they are often considered a new state of matter. The diverse Behavior of granular materials makes it extremely difficult to establish a general theory that accounts for the observed phenomena. Some fundamentals have been revealed. For example, when the average energy of the individual particles is low and the particles are fairly stationary relative to each other, the granular material acts like a solid. In general, stress in a granular solid is not distributed uniformly but is conducted away along force chains. When the granular matter is driven and energy is fed into the system such that the particles are not in constant contact with each other, the granular material typically enters a liquid-like state. When freely flowing granular materials have flow characteristics that roughly resemble those of ordinary Newtonian fluids. However, granular materials dissipate energy quickly, so techniques of statistical mechanics that assume conservation of energy are of limited use. Bulk flow characteristics of granular materials do differ from those of homogeneous fluids. If the granular material is driven harder such that contacts between the particles become highly infrequent, the material enters a gaseous state. Unlike conventional gases, granular gas will tend to cluster and clump due to the dissipative nature of the collisions between particles.

1.2 Static behaviors

Owing to the dissipative interactions between the particles, a granular assembly without energy input is at rest in a metastable state, and the typical static properties include Coulomb friction, the silo effect, effective stress, the Rowe stress dilatancy relation, and so on.

1.2.1 Coulomb friction law

In 1773, French physicist Coulomb regarded the yield of granular matter as a friction process and proposed the friction law, that is, the friction of solid particles is proportional to the normal pressure between them and the static friction coefficient is greater than the sliding friction coefficient. At that time, he was looking for the macroscopic criteria for soil failure and was not interested in the soil mass movement Behavior after soil collapse, but this is precisely of concern in granular flow.

The Coulomb also studied the phenomenon of collapse in piles of sand and found that when the accumulation angle was greater than a certain value the collapse occurs. The sandpile collapse is different from liquid flow, as only those particles within a thickness of about 10 particles on the sandpile's surface fall down, while the deeper particles are static. It is found that when the sandpile's slope reaches the maximum stable angle, the sand particles on the surface begin to fall and stop until the surface inclination angle is equal to the angle of repose, that is, only the particles between the angle of repose and the maximum stable angle on the sandpile's surface collapse. It is well established that any experimental measurement of the macroscopic angle of tilt lies between two limits, the maximum angle of stability θ_m and the minimum angle of repose θ_r; the difference between these two angles $\Delta\theta = \theta_m - \theta_r$ is the so-called Bagnold angle ($\sim 2°$), which is a measure of the hysteresis of granular materials.

The surface roughness and size distribution can affect the angle of repose. The typical value of the angle of repose of sand is 35°. Smooth particles correspond to a small angle of repose. This phenomenon can be qualitatively explained by force chains. Under the action of gravity, the force chains are formed parallel to the direction of the contact forces due to extrusion, in order to effectively transit the gravity. The spherical particles have a slight deformation at contact points, which can support small shear stresses. When the shear stress exceeds a certain value, the force chains tend to break and the particles at the end of force chains on the sandpile's surface begin to slide. However, new force chains continue forming until the shear stress is less than the critical value, and so the sandpile maintains a constant angle of repose.

The piles of sand may be resulted from different construction techniques, and there are obvious differences among the force chains and the stress distributions within the piles. Starting with a uniform density, the pressure at the bottom of the pile is found to show a single central peak. It turns into a pressure dip, if some density in homogeneity, with the

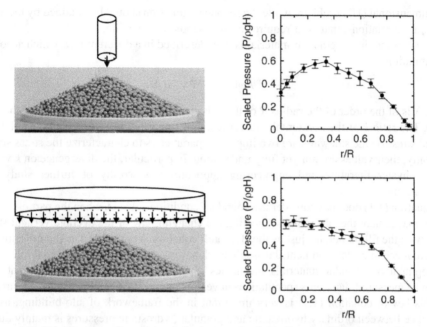

Figure 1.3: Distribution of pressure at the bottom by funneling (upper) and rain-like pouring (lower).

center being less compact, is assumed. These two pressure distributions are remarkably similar to recent measurements, in the form of piles obtained, respectively, by rain-like pouring and funneling, as shown in Figure 1.3. It indicates that the construction history affects the pressure distribution at the bottom of sandpiles formed on a rigid base. A heuristic explanation of the mechanisms producing the dip is that the flow of particles during the funneling procedure forms stress chains oriented preferentially in the direction of the slope. These chains form arches that shield the center from some of the weight, thereby forming the dip.

1.2.2 Janssen effect

In 1895, Janssen discovered that in a vertical cylinder filled with particles, the pressure measured at the bottom does not depend on the height of the filling, that is, it does not follow the Stevin law that is valid for Newtonian fluids at rest. The first interpretation of the law has been provided by Janssen, in terms of a simplified model with the following assumptions: (1) the vertical pressure σ_{zz} is constant in the horizontal plane; (2) the horizontal pressure σ_{rr} is proportional to the vertical pressure σ_{zz}, where $K = \sigma_{rr}/\sigma_{zz}$ is constant in space; (3) the wall friction $\mu = \sigma_{rz}/\sigma_{rr}$ (where μ is the static friction coefficient) sustains the vertical load at the contact with the wall; and (4) the density of the material is constant over all depths.

In particular, the first assumption is not true since the pressure also depends on the distance from the central axis of the cylinder, but this is not essential as it is formulated as a one-dimensional (1D) problem, while the second assumption should be obtained by means of constitutive relations, that is, it requires a microscopic justification.

The pressure in the granular material is then described in a different law, which accounts for saturation:

$$p(z) = p_\infty[1 - \exp(-z/\lambda)] \tag{1.1}$$

where λ is of the order of the radius R of the cylinder and $\lambda = R/2\mu k$. At the top of the silo, $z = 0$. The ratio k of the horizontal and vertical stress and the ratio μ of the tangential and normal stress on the sidewalls are two important parameters to characterize the stress status. Currently, their values are not yet fully understood. In particular, the divergence on k values in the physics literature and engineering applications is worthy of further study and clarification.

Equation (1.1) does not consider the boundary conditions, that is, when the particle size is much smaller than the silo radius, reflecting the changing law of the average vertical stress along with the filling height; just as the Bernoulli equation is considered as the milestone of the fluid mechanics development, Janssen's work has played an extremely crucial role in the development of granular materials mechanics. Since the internal stress of the materials cannot be measured, Janssen's speculations have not been verified by any direct experiments.

This law of equation (1.1) is very important in the framework of silo-building, as the difference between ordinary hydrostatic and granular hydrostatic pressures is mainly due to the presence of anomalous side pressure, that is, forces exerted against the walls of the cylinder. It happens that the use of a fluid-like estimate of the horizontal and vertical pressures leads to an underestimation of the side pressure and, consequently, to unexpected explosions of silos. It can be seen from equation (1.1) that near the free surface of the assembly, $z \ll \lambda$, the internal pressure P is line with the hydrostatic pressure, $P \propto z$; when near the bottom of the silo, $z \gg \lambda$, $P \approx P\infty$ (see Figure 1.4).

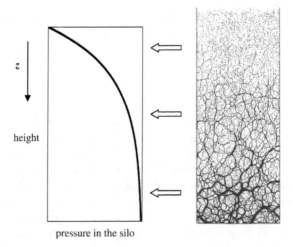

pressure in the silo

Figure 1.4: Pressure distribution in a silo and a schematic diagram of internal force networks.

From the force chain directions shown in Figure 1.4, it can be seen that a fraction of the gravity is transmitted to the side wall. As close to the bottom of the silo as possible, force chains fully connect with the side wall so that the greater part of the gravity would transmit to the side wall as the depth increases, which makes the bottom pressure no longer change with the increase in depth of granular matter.

1.2.3 Effective stress

Since the 1920s, engineering failures and accidents often occurred, such as the saturated sand liquefaction of railway embankments. Terzaghi first proposed the relationship for effective stress in 1936. The term "effective" referred to the calculated stress that was effective in moving soil or in causing displacements. It represents the average stress carried by the soil skeleton, and it is calculated from two parameters, namely total stress and pore water pressure, while the shear strength of the soil is determined by the effective stress.

Along with the introduction of the concept of effective stress, the study of soil mechanics began separating from applied mechanics as an independent discipline. We can see that the important development stages of soil mechanics in the past century were all closely related to the deep understanding of the properties of constituent particles, new experimental techniques, and theoretical breakthroughs at particle scales. For example, the effective stress principle in soil mechanics is exactly the same as the Archimedes buoyancy principle in physics.

A saturated soil is a relatively simple case, as shown in Figure 1.5. A vertical loading P acts on the horizontal total area A. P is shared by the component force P' of the contact between the two particles and the hydrostatic pressure $(A - A_c)u$.

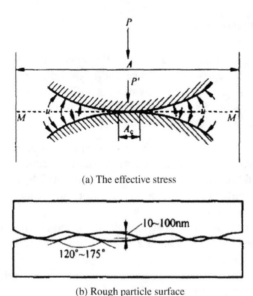

(a) The effective stress

(b) Rough particle surface

Figure 1.5: A schematic diagram of the loading transmission. The cross-section is chosen on the contact points between particles.

$$P = P' + (A - A_c)u \tag{1.2}$$

where A_c is the contact area between the particles. Both sides in equation (1.2) are divided by the total area A, after which we have:

$$\frac{P}{A} = \frac{P'}{A} + \left(\frac{A - A_c}{A}\right)u \tag{1.3}$$

It can also be expressed as:

$$\sigma = \sigma' + \left(1 - \frac{A_c}{A}\right)u \tag{1.4}$$

The contact area A_c is very small. Sometimes, it is usually considered as a point contact. The stress at the contact point often reaches the yield stress of materials in reality:

$$A_c = \frac{P'}{\sigma_y} \tag{1.5}$$

where σ_y is the yield strength of the constituent particles.

Under normal circumstances, A_c is negligible. Equation (1.4) can be approximated as:

$$\sigma = \sigma' + u \tag{1.6}$$

This is the expression of effective stress in saturated soils that was first proposed by Terzaghi. It can be seen from equation (1.6) that the effective stress σ is a nominal physical quantity, which is the sum of the horizontal cross-sectional areas divided by the sum of the vertical components of the contact forces between the soil particles. Therefore, it is much smaller than the actual contact stress.

1.2.4 Rowe stress–dilatancy relation

The bulk properties of granular materials actually reflect the collective Behaviors of the constituent particles. In the early 1960s, Rowe studied the influence of dilatancy on the shear strength in shear tests and proposed the famous stress–dilatancy relation. However, the study of the microscopic particle arrangements in terms of macroscopic mechanical responses requires the consideration of the following four issues:

(1) A need to establish the relationship between the stress tensor and the interparticle force on the basis of force balance analysis, corresponding to equation (1.7).
(2) Obtaining of the deformation coordination equation, which should be satisfied during particle displacements, by considering the structural characteristics of the RVE, corresponding to equation (1.8).
(3) Establishment of the relationship between the force and the displacement of the contact points, that is, when the friction angle reaches the maximum friction angle, the contact points transition to relative sliding.
(4) Transformation of interparticle contacts to stress–strain relations based on the correlations among equations (1.7), (1.8), and (1.9).

These procedures are essential to establish a macroscopic stress–strain relation. In equations (1.7), (1.8), and (1.9), both stress and strain have been endowed upon the interparticle forces, contacts displacements, and particle movements. Rowe used a minimum energy ratio principle to develop this relation, and later it was derived again by considering toothed separation planes.

Corresponding to the regular packing of mono-dispersed particles in 2D in Figure 1.6, the mechanical properties of the assembly can be analyzed based on the RVE, shown as the shaded area. The typical lengths in vertical and horizontal directions of the RVE are l_1 and l_2, respectively. The variable $\alpha = \arctan(l_1/l_2)$ describes the geometric characteristics of the RVE. The variable β represents the angle when the contact points begin the process of sliding. The vertical direction σ_{11} is the direction of the major principal stress, and the horizontal direction σ_{22} is the direction of the minor principal stress.

Stress on the boundary can be expressed as the concentrated force borne by individual particles. Under biaxial loading with uniform stress, $\sigma_{12} = \sigma_{21} = 0$, that is, $F_{12} = F_{21} = 0$. At equilibrium state, F_{11} and F_{22} acting on the RVE satisfy the following equation:

$$\frac{F_{11}}{F_{22}} = \frac{\sigma_{11}l_2}{\sigma_{22}l_1} = \tan(\theta + \beta) \qquad (1.7)$$

where θ is the so-called actual friction angle, that is, the angle between the contact force and the contact normal direction. For a static granular system, both contact force and torque are balanced. Particle surface friction enhances the capacity of a granular system to maintain stability under external loadings. If the particle surfaces are smooth, and if the resultant force is perpendicular to the contact surface, the particles are able to remain steady; if the particle surfaces are rough, as long as the tangential force falls within the scope of the friction cone θ_μ, the particles would still remain steady. θ_μ is determined by the coefficient of friction $\mu = tg\phi_u$, so $\theta \leq \theta_\mu$.

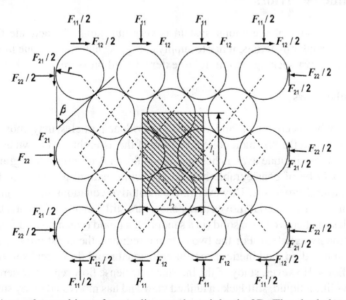

Figure 1.6: A regular packing of monodispersed particles in 2D. The shaded areas represent the representative volume elements.

When the applied deviatoric stress causes θ to gradually increase the limit value θ_μ, while maintaining the variables α and β unchanged, the internal structure will not evolve until the two vertical particles come into contact, and the microstructure is re-arranged. Once $\theta \geq \theta_\mu$, the particles will commence the act of relative sliding, resulting in change in the element structure and creation of new force chains. Then, the displacement in horizontal and vertical meets the following equation:

$$\frac{\dot{\Delta}_2}{\dot{\Delta}_1} = \frac{\dot{\varepsilon}_{22} l_2}{\dot{\varepsilon}_{11} l_1} = -\tan \beta \tag{1.8}$$

That is,

$$-\frac{d\dot{V}}{V\dot{\varepsilon}_1} = 1 - \tan \alpha \tan \beta \tag{1.9}$$

The widely used stress–dilatancy equation is:

$$\frac{\sigma_{11}}{\sigma_{22}\left(1 + d\dot{V}/V\dot{\varepsilon}_1\right)} = \tan^2\left(45 + \frac{\phi_u}{2}\right) \tag{1.10}$$

The Rowe stress–dilatancy relation establishes the relationship of the macro-shear strength with the particle friction coefficient ($\mu = tg\theta_u$) and dilatancy. For noncohesive granular materials, the most important parameter to determine the strength is the friction coefficient, which is verified by a large number of triaxial tests on dense granular materials.

1.3 Dynamic behaviors

The dynamic behavior of a granular system is very intriguing. To activate the motion of initially static granular materials, it has to supply energy, which can be done in various ways, such as shearing, avalanching, vibrating, or external fluid pressure.

1.3.1 Granular flows

Subjected to unbalanced forces and torques, individual particles may move by obeying Newtonian laws, and the resultant collective flow of the assembly is known as the so-called granular flow. These include avalanches and debris flows. A description of granular flows in continuum would be of considerable help in predicting natural geophysical hazards or in designing industrial processes. However, the constitutive equations for dry granular flows, which govern how the material moves under shear, are still a matter of debate. One difficulty is that particles can behave like a solid (in a sandpile), a liquid (when poured from a silo), or a gas (when strongly agitated). For the two extreme regimes, the constitutive equations have been proposed based on the kinetic theory of collisional rapid flows, and soil mechanics for slow plastic flows. However, study of the intermediate dense flow regime, where the granular material flows like a liquid, still lacks a unified view and has motivated many studies over the past decade. The main characteristics of granular liquids include a yield criterion (a critical shear stress below which flow is not possible) and a complex dependence on shear rate when flowing. In this sense, granular matter shares similarities with classical viscoplastic fluids such as Bingham fluids.

Granular flows are generally divided into quasi-static flows, slow flows, and rapid flows. In the quasi-static flows, the force network continues to exist, in general, and the macroscopic shear stress $\tau \propto \gamma$, where γ is the shear strain. Some elastic–plastic constitutive relations have been proposed for quasi-static flows. In slow granular flows, the force network frequently forms and self-deconstructs, while the shear stress $\tau \propto \dot{\gamma}$, where $\dot{\gamma}$ is the strain rate. As the flow rate increases such that the system attains a state typical of rapid flows, the correlations between the stress and strain rate increase until $\tau \propto \dot{\gamma}^2$. The first $\dot{\gamma}$ represents the momentum exchange between particles, and the second $\dot{\gamma}$ reflects the collision frequency.

In 1954, Bagnold published his findings on the rheological properties of a liquid–solid suspension. In his experiments, paraffin and stearic acid leads were mixed to create particles of 0.132 cm diameters; the density of these particles is almost equal to the density of water, so the particles are suspended in the water. A sketch of the Bagnold experimental apparatus is shown in Figure 1.7. The radii of the outer and inner cylinders were $r_o = 5.7$ cm and

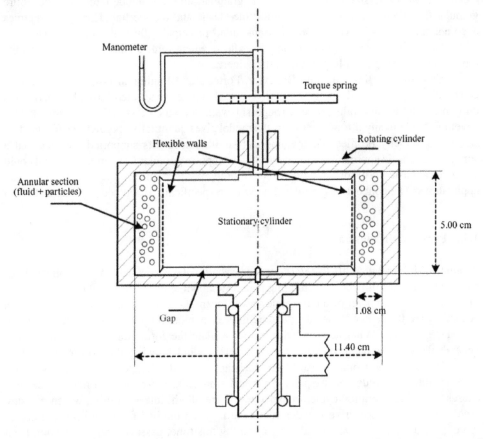

Figure 1.7: Experimental apparatus used by Bagnold [5] to measure the shear and normal forces in a sheared suspension.

$r_i = 4.62$ cm. The height h of the test section was 5 cm. The outer cylinder of the rheometer was rotated at a constant rate ranging from 15 to approximately 500 r.p.m. The inner cylinder was held fixed using a spring that enabled a measurement of the torque. The momentum transfers between suspended particles are realized from frequent collisions; the corresponding stress is known as the Bagnold stress. For the rapid granular flow, the concept of granular temperature is introduced, and the kinetic theory of particles is used for theoretical research.

The interparticle forces are not evenly distributed throughout the material, but are concentrated in force chains. These are quasi-linear structures that support the bulk of the internal stress within the material. In a shearing material, these force chains are dynamic structures. When the material shears, particles are pushed together to form the chains. After a chain is formed, it will be rotated slightly by the shear motion, but will quickly become unstable and collapse. If the solid concentration is small or the particle surfaces are smooth, stable force chains cannot be formed, and there occur frequent collisions between particles. This is a way of transferring force or momentum changes in granular flows. Based on whether there exist stable force chains in granular flows, Campbell divided the entire granular flow field into two broad regimes: the elastic and the inertial. The elastic regimes encompass all flows in which force is transmitted principally through the deformation of force chains. The inertial regime encompasses flows where force chains cannot form and the momentum is transported largely by particle inertia.

Inspired by the Bingham fluid Behavior, Forterre and Pouliquen [18] proposed a new constitutive relation for dense granular flows. A 3D model was tested through experiments on granular flows on a pile between rough sidewalls, in which a complex 3D flow pattern develops. Without any fitting parameter, the model gives quantitative predictions for the flow shape and velocity profiles. Their numerical simulation results supported the idea that a simple viscoplastic approach could quantitatively capture granular flow properties, and could serve as a basic tool for modeling more complex flows in geophysical or industrial applications. The concept of granular flow will be specifically discussed in Chapter 8.

1.3.2 Faraday circulation

Around 1831, Faraday (1791 to 1867) reported a shallow layer of a fine powder collected in a circular heap at an antinode of a vibrating plate. As shown in Figure 1.8, granular flow occurs in a shallow surface layer of the heap, downward from its center to its edge; inward motion occurs toward the center at the edge; and by inference, motion upward toward the middle of the heap's interior. What he saw led him to postulate the formation of a partial vacuum beneath the powder heap, creating (1) an external wind blowing the powder toward an antinode and (2) a flow of air inward from the heap's edge, driving powder toward its center. At a suitable amplitude and frequency, vertical sinusoidal vibration of a fine-powder bed causes it, in each vibration cycle, to experience a free-flight interval during which pressure gradients in its interior drive powder centerward. When the bed-floor collision terminates flight, pressure gradients reverse direction; but by this time, passage of a compaction front has locked particles against further movement. Before the next flight interval, an increase in porosity will reverse the compaction that accompanied the bed-floor collision.

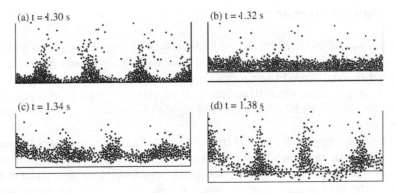

(a) t = 1.30 s
(b) t = 1.32 s
(c) t = 1.34 s
(d) t = 1.38 s

Figure 1.8: Granular circulations when the system is being vertically vibrated.

1.3.3 Reynolds' dilatancy

Reynolds' dilatancy is the observed tendency of a compacted granular material to dilate (expand in volume) as it is sheared. This occurs because the particles in a compacted state are interlocking and therefore do not have the freedom to move around one another. As illustrated in Figure 1.9, when stressed, a lever motion occurs between neighboring particles, which produces a bulk expansion of the material. On the other hand, when a granular material begins in a very loose state, it may initially compact instead of dilating under shear. Reynolds' dilatancy is a common feature observed in the soil and sand systems studied by geotechnical engineers, and this is a part of the broader topic of soil mechanics.

It is well known that when we walk on wet sand on the shore, the footstep impressions become dry. This is similar to what occurs in our setup. When a deformation is imposed on the sand, the spaces between particles increase, allowing for upper layers of water to invade the sand. As a consequence, the footstep impressions become dry. It was first described scientifically by Osborne Reynolds (1842 to 1912) in 1885 [21] and 1886.

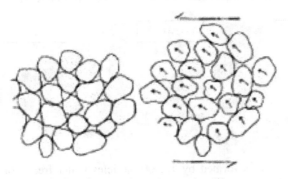

Figure 1.9: Illustration of granular dilatancy.

1.3.4 Clustering in granular gas

Granular gas may be clustered because of its dissipative nature. For a group of particles with uniform spatial distribution, they may form a cluster so tightly packed that there occur an infinite number of collisions in a finite amount of time. This process is the so-called inelastic collapse. As long as there is insufficient energy input to counter this dissipation, the granular gas will form clusters with high density, making the system a serious nonuniform density distribution.

For a free-cooling granular gas, where there is no external driving force, the velocity of particles movement decreases due to dissipation. The process is shown in Figure 1.10. Theoretically, if this process continues, the velocity distribution and the spatial distribution both tend to the form of the δ function, where the overall velocity distribution approaches zero, and the spatial distribution will not arrive at a point due to the volume exclusion effect.

For the wall vibration-driven granular gas, the nonuniformity of density is accordingly produced along the driving direction. If there is not enough wall energy input to balance the energy loss of the system, the particles will move to the direction away from the vibrated walls, making the density nonuniformly expanded. The particles will eventually form clusters away from the vibrated walls.

1.4 Force measurement and internal structure recognition

It is typical to rely on boundary measurements performed during actual prototype monitoring or on physical model tests in the laboratory. However, the advances in granular materials are greatly dependent on the developments in related physics and mechanics, measurements of interparticle forces, and the recognition of internal packing structures. There are only a few nondestructive techniques available to observe and quantify the patterns of local deformation of granular materials, such as X-ray radiography and photoelastic stress analysis.

1.4.1 X-ray radiography

Radiography uses penetrating radiation, such as X-rays, to produce shadow images of the internal structure of materials. Since the discovery of the principle by Roentgen in 1895, the

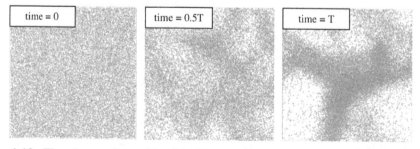

Figure 1.10: The clusters formed by 10,000 particles under free cooling. The restitution coefficient is 0.5.

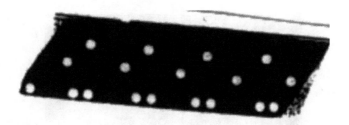

Figure 1.11: Overlay of radiation images of deformation fields in a sand sample during shear
            experiments.

application of radiography in the earth sciences and soil engineering has been slow to
develop. However, recent work shows that it has great potential.

To investigate a granular assembly, it is possible to use X-ray irradiation to penetrate the
material and record images of their internal state. With X-ray imaging, the internal state of
the granular materials subjected to boundary loads can be determined quantitatively, while
minimizing both disturbance and boundary effects on the observed system. Small lead shots
placed in the model can be used as displacements markers for granular materials. The
information that has been obtained on the deformation mechanisms shows that X-ray
imaging techniques have the potential to improve our understanding of granular materials. In
the early 1960s, Roscoe demonstrated the movement of the lead shots under simple shear,
with the X-ray radiation imaging technique. Figure 1.11 depicts the overlying picture before
and after a strain increment, after the system has reached the peak stress (elongated box). The
white dots represent the lead particles. If the shear deformation is uniform, the white lead
particles in the middle and bottom of this picture should be visible. However, there are only
two visible images of lead particles in the lower layers in the superimposed picture,
indicating that shear deformation is concentrated in the strip between the middle and the
lower layers of lead particles, that is, the direction of the failure surface is horizontal without
any elongated direction. It confirms that the direction of the failure surface (relative to the
principal stress direction) is not the direction of Mohr–Coulomb failure surface, as it
deviates from $45 \pm \varphi/2$, instead tending toward $45 \pm \psi/2$. The variables φ and ψ represent
the internal friction angle and dilatancy angle of sand; in general, they are significantly
different. This was further verified in the tests of 1 g scale models in centrifuge tests. Soil
mechanics researchers in Cambridge University have developed the critical density concept
of sand, proposed earlier by Casagrande, into a critical state applicable for sands and clay,
and have linked the shear deformation behavior of a soil mass with its change in volume by
the original Cam-Clay model and the Modified Cam-Clay model. These represent the so-
called critical state soil mechanics.

Since the 1970s, experimental methods have developed greatly, mainly benefiting from
the development of computer and information technology, and microdestructive testing
techniques. X-ray CT techniques are gradually being used in the study of internal structures
and their evolution, and study of granular matter by scanning and image reconstruction.
Figure 1.12 shows the evolution of porosity in the 3D granular system, in which a shear band
can be clearly observed.

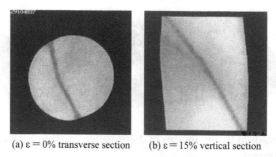

(a) $\varepsilon = 0\%$ transverse section　　(b) $\varepsilon = 15\%$ vertical section

Figure 1.12:　X-ray CT images of a shear band in a granular assembly.

It is noteworthy that a simple shear test initiates the principal stress rotation; the test results are more general than other tests, such as the plane compression test or the triaxial compression test. Although the hollow cylinder rotation apparatus also allows for the principal stress rotation, the operations are too complex, and granular sample preparations are cumbersome. It was abandoned early. During that period, the Cambridge researchers also developed a method utilizing scanning electron microscopy of clay to study the structural and mechanical properties of clay, especially the shear strength anisotropy. The contribution of Cambridge researchers to soil mechanics was to introduce new experimental methods and employ a relatively "simple" model to represent the principles of soil mechanics, in order for it to be learned by younger scholars.

1.4.2　Photoelastic stress analysis

Photoelasticity is a stress analysis method using optical principles. Its basic principle is to place a sample made from transparent plastic with stress birefringence effect (e.g. epoxy resin plastic and polycarbonate plastic) in a polarized light field. At this time, the loading is imposed on the sample, following which the interference fringes can be seen on the model; the stress distribution of a structural model under loading can be observed by measuring the interference fringes.

The first to use photoelastic experiments in the study of granular materials were civil engineering researchers. In 1957, Dantu [13] and Wakabayashi [2], respectively, observed the force transmission in granular materials by using photoelastic techniques. Through the experiments, a clear optical band was observed that represents the major principal stress direction. However, the experiment was only used for qualitative analysis, and not for quantitative analysis of stress distribution. De Josselin de Jong and Verruijt [7] simulated 2D granular materials with cylindrical photoelastic disks, and quantitatively described interparticle forces by manually demarcating the isochromatic fringe orders at particle contacts. Afterwards, by using this method and combining the established average method, the second-order stress strain tensor could be obtained. Drecher and De Josselin de Jong [8] further studied the stress rotation. Their results showed that the second-order tensor is effective in describing the contact force distribution within a certain region. At the same time, Oda et al. [9] studied the contact force distribution and structure changes in 2D granular materials by using columnar photoelastic disks, as well as the anisotropy of a granular assembly by using oval-shaped photoelastic cylinders. Although the particle shapes are mostly depicted to be cylindrical or elliptically cylindrical, they are very different from the actual shapes of granular materials, which make applications of these experimental methods limited.

To better simulate the mechanical Behavior of actual granular materials, Drescher [10] and Allersma [11] selected broken glass as photoelastic granular material. In particular, Allersma improved Drescher's experimental methods by rotating the analyzer point-by-point to measure the light intensity and indirectly measure the phase difference of specific points inside the granular material. Accordingly, the stress distribution fields and the displacement distribution fields within granular materials were obtained. The development of photoelastic granular experimental mechanics not only promoted the basic theory of granular mechanics and plasticity mechanics, but it also gave birth to the Particle Flow Code program (PFC) which generated huge influence afterwards.

In the past 20 years, with the popularity of monochrome polarized light sources, the emergence of the high-precision digital camera, and the development of digital image processing techniques, the "old" photoelastic method has played an important role in the display and analysis of the nonuniformity of stress distribution within granular systems, as well as understanding of the complex collective mechanical Behavior of granular systems. Researchers from Duke University conducted detailed photoelastic granular experiments, analyzed the average stress within the particles by the gray-level gradient G2 of granular photoelastic images, and obtained the contact force distribution and force chain structure morphology of particles under point loading and direct shear. Since 2007, Tsinghua University has been conducting photoelastic experiments of granular systems. Figure 1.13 shows the photoelastic fringes of a single particle. The larger the pressure, and the higher the fringe order within the particles, the larger the gray gradient.

The earliest physical experiment was the penetration test, that is, the use of a probe or core sampler to penetrate static or dynamic forces in granular materials, so as to obtain physical mechanical properties within the granular materials, as shown in Figure 1.14.

It is necessary to adopt 2D photoelastic tests to study the granular mechanics at this stage, since 3D measurements cannot be effectively conducted so far. As a noninvasive method of detection, it does not cause any disturbance to the granular system. It can not only measure

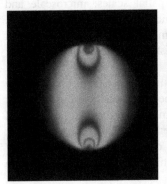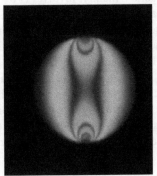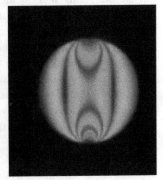

Figure 1.13: Stress stripes within particles at different pressures and calculation diagram of G^2.

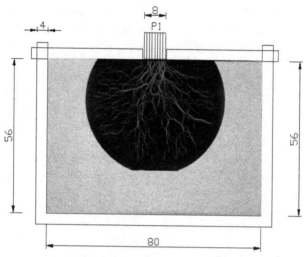

Figure 1.14: Static penetration tests with photoelastic disks. The force chain network can be clearly observed in the polaroid coverage area.

the interparticle forces, but can also judge the positions, and then clearly observe the force chains. However, the disadvantage of this method is that the measurement accuracy of contact force is not high.

1.5 Granular physics

Granular materials are large conglomerations of discrete macroscopic particles. The study of granular materials is of great engineering importance, and innumerable empiric equations have been presented to fit otherwise meaningless data. However, nonlinear mechanical Behaviors are still rather poorly understood. From the perspective of the structure of matter, the simple fluids and ideal solids mainly involve two spatial scales: microscale and macroscale, which obey quantum mechanics and classical mechanics, respectively. Meanwhile, the thermodynamic properties of a system with macroscopic scale matter could be directly derived from the microscopic matter by using statistical physics or kinetic theory; thus, the correlation between the two scales is established.

Each particle is of course a solid and can be described with elastoplastic mechanics. The collective behavior of all constituent particles can be described with classical mechanics as well, but such systems behave substantially differently from ordinary solids, liquids, and gases. The key problem is that the macroscopic parameters of granular materials (e.g. friction, elasticity, plasticity, and so on) cannot be simply derived from the statistics of particles' positions, velocities, and the interparticle forces. In other words, this challenges the application of statistical physics.

In granular materials, the kinetic energy and potential energy (due to gravity and elastic deformation) become of paramount importance. Because $k_B T$ in granular materials is near zero, unless excited by external disturbances, each metastable configuration of the material will last indefinitely, and no thermal averaging over nearby configurations will take place. The granular system is not a conservative system; even if the particles undergo motion by external disturbances, the particles' surface friction and inelastic collisions will also lead to dissipation of kinetic energy. For these reasons, granular matter is nonergodic; this has destroyed the foundation of statistical physics.

From the mechanical point of view, granular systems show complex Behaviors, which are different from ordinary liquids and elastic solids. These Behaviors include strong nonlinearity and irreversible deformation, solid-like and liquid-like transitions, unique rheological properties of granular flows, and so on. Over 20 years of theory development, the study of granular material has been developed into the leading edge of physics and mechanics.

1.5.1 Granular solid hydrodynamics

The macrokinetics properties of common liquids and solids can be described separately by the Navier–Stokes equation and the elastic equation. But the mechanical phenomena of granular matter are more complex; the stress–strain relation is nonlinear and irreversible and shows Behavior common to both fluids and solids. Applications in engineering and physics in the past 20 years have made many measurements on these phenomena, and have accumulated a large amount of experimental data and empirical formulae. How to build a quantitative description of their dynamics is a basic research topic with important application value.

Geometric equations, thermodynamic relations, and the various conservation laws are classical physics bases to deal with the macroscopic mechanical properties of nonuniform objects. They have successfully established the continuum mechanics theory of common liquids and solids using the energy, momentum, mass density, entropy, strain, temperature, stress, increase rate of entropy, and a variety of dissipative flows. Whether the classical physical methods can handle the study of granular matter not only indicates the promotion and extension of their scope, but also demonstrates a serious challenge to their ability. According to the classical physics, continuum mechanics theory of any material is composed of general equations and material equations; the former refers to the conservation equation, the entropy increase equation, thermodynamic identities and Onsager relations, while the latter is the expression of the thermodynamic characteristic function and the migration coefficient. Physics often regards this kind of continuous medium theory established on the basis of classical physics as "hydrodynamics." The name only expresses that their concepts are from Newtonian fluid mechanics, rather than indicating that the materials described must be a liquid or gas. Examples include liquid crystal fluid dynamics, polymer fluid dynamics, crystal fluid dynamics, and so on.

As is well known, the expression form of the disordered motion of micromolecules at macro-level is called thermodynamic entropy or temperature, but the expression form of the meso-level motions at the macro-level is still relatively unfamiliar, and is a new research topic in physics. If the ordered crystalline phenomenon occurring in the spherical granular system with the same radius is ignored, the particle structure and motion at meso-levels include disorder components without long-range correlations, which can be classified as a class of motions with random fluctuations. This point is very similar to the thermal motion of molecules. However, apart from this similarity, there is an important difference between the disordered motion at meso-level and thermal motion of micromolecules. Since the interaction between particles is normally inelastic, when the moving particles are not in a state of local equilibrium, there is entropy production. In other words, as long as the molecular motion of a general gas reaches the Maxwell velocity distribution, the gas is in the thermodynamic equilibrium state; even if nonstop collisions between molecules exist, there is no entropy production. When the moving granular system is excited to the Maxwell velocity distribution state, the carried kinetic energy will reduce or relax in the form of microscopic molecular motion, and this is always accompanied by entropy production. It needs to be specifically noted that entropy production is a process with limited time, which means that the disordered motion at the meso-level cannot be directly attributed to the thermodynamic entropy. Therefore, an additional entropy-like variable s_g, to reflect its existence and evolution, has necessarily to be introduced.

The above-mentioned meso-characteristics are the main differences between granular matter and ordinary elastic objects, which force two types of materials to exhibit significantly different mechanical Behaviors. Jiang and Liu [14] have tried to extend the Newtonian fluids and fluid dynamics theory of elastic materials to granular matter in order to establish the macro-kinetic theory of granular matter, "Granular Solid Hydrodynamics" (GSH). To reflect the features of "disorder" and "non-equilibrium," GSH theory has the newly introduced variable s_g with entropy-like Behavior and relaxation; this is in order to distinguish it from the thermodynamic entropy s; s_g is called granular entropy. GSH adheres to the basis of classical physics and uses the thermodynamics characteristic function and migration coefficient expressions for modeling objects, which are the main features of fluid dynamics of granular matter. These indicate the reason for why it is so different from engineering constitutive theory, in terms of methods and concepts.

Some widely used theories on the constitutive relations of granular matter often do not comply with all the above-mentioned classical physics, as they cannot completely describe the physical properties and can only be considered as empirical formulae to describe some aspects of experimental data. Under a reasonable approximation, GSH theory should be able to explain the empirical formula and generate the premise of their application and limitations. Since the experimental data on granular matter is particularly complex and diverse, the research on the material models of granular matter is still in its infancy. GSH theory still has a limited understanding of those materials models, especially the migration coefficient. It mainly refers to the materials without adhesion between particles, such as dry sand and others. With a little encouragement, these models can also be used to describe the materials with adhesion interactions between particles, such as wet sand and so on. Faced with various complex factors, such as nonlinearity, plasticity, dissipation, and transition Behavior across the fluid, solid etc., whether the classical physics framework can successfully solve the macromechanical basis of such materials is also a worthy scientific problem to be studied and investigated. Since there is no reason, in principle, to believe that granular matter would behave contrary to the principles of classical physics, this research will be mainly reflected in the research and improvement of material models, as well as in the comparative discussions of theoretical calculations and experimental data. From the current work, GSH results are quite satisfactory.

1.5.2 Statistical mechanics

Granular materials are essential multibody systems consisting of discrete solid particles. In spite of a simple nature of interparticle interactions, the collective Behavior of such systems is still poorly understood. This is similar to thermal systems. For example, under specified conditions granular materials exhibit ordered patterns, such as ripples in sand. The occurrences of such kinds of ordered patterns are the emergent phenomena, which is the fingerprint. The analogy to thermal systems raises a question. Could statistical mechanics be applied to granular matter?

Statistical mechanics of granular matter is a new and rapidly developing subset of classical statistical mechanics. However, there exists no theoretical framework which can explain the observations in a unified manner beyond the phenomenological jamming diagram. Many physicists believe that granular matter still obeys some fundamental principles of statistical mechanics, and some successes have been made in the past years. For example, Edwards' generalization of statistical mechanics has mostly been applied within the context of compaction, where the main observable variable is volume, so that the spatial arrangement of particles matter. Yet, if the forces between particles are also taken into account, a similar approach can be used to statistically describe the mechanical properties of static granular matter.

1.5.3 Granular gas

If the granular material is driven even harder, such that particle collisions become highly frequent, the material enters a gaseous state. Granular gases are rarified systems of particles that freely move between dissipative collisions. Numerous surprising effects exist, such as anomalous diffusion, violation of energy conservation, existence of different temperatures for the translational and rotational motion of particles, correlation of spin and velocity, and spontaneous formation of shocks, vortexes, and clusters in initially homogeneous systems. These, as well as some other effects, have been carefully studied so far. In granular gases, the assumptions of continuity and binary collisions need to be considered carefully; the inevitable individual differences among particles make it impossible for the corresponding experimental studies to be precisely repeated, while the seemingly simple interactions between individuals are difficult to handle by mathematics. This causes difficulties in theoretical descriptions of granular gases. The relevant research still attracts widespread attention because granular gas possesses the basic characteristics of multidissipative systems; the research into it will help in understanding nonequilibrium systems and nonlinear phenomena.

Strictly speaking, there is no clear definition of granular gas. Analogous to the classical molecular gas, when the spatial distribution of particles in the system is dispersed, the average free path of particle movement is much larger than the particle size. Furthermore, the collisions between particles are mainly binary collisions, and so the system can be considered as a granular gas. Granular gases and classical molecular gases can be simply understood as hard-sphere systems, but there are fundamental differences between the two, as discussed in the following list.

(1) *Identity of particles:* Molecules can be considered as identical, but there are no two identical particles. This difference is reflected in many aspects, including size, shape, mass, surface roughness, and nonelastic degree, which causes some random effects. It is generally assumed that all the particles have the same physical characteristics, and such assumptions have no great influence on the physical Behavior of the granular gases. However, the fact that particles are not identical has important consequences in statistical mechanics. Calculations in statistical mechanics rely on probabilistic arguments, which are sensitive to whether or not the objects being studied are identical. As a result, identical particles exhibit markedly different statistical Behavior from distinguishable particles.

(2) *Energy conservation:* Although molecular gas can be described by the classical collision model, the interaction between molecules is essentially of a quantum mechanical nature, which is generally considered to involve no energy loss. The collisions between particles involve plastic deformation and frictional dissipation. These kinds of dissipative characteristics make the granular gas show different characteristics from the molecular gas.

(3) *Nonhead-on collisions:* van der Waals interactions between microscopic molecules within a particle can be ignored for the macroscopic motion of a particle. In a granular gas, the contact force is the primary interaction between particles, which belongs to a short-range interaction on the order of the particle diameter. This is because most collisions are not *head-on* collisions, leading to *head-on* interaction between particles. The tangential friction force has strong nonlinear characteristics and makes the particles rotate. Therefore, it is more difficult to describe a granular gas a molecular gas.

(4) *Binary collisions:* It is usually assumed that particle distributions in both granular gases and classical molecular gases are sparse enough that the collisions between discrete individuals are binary. For molecular gases, it is reasonable, but for the granular gases with a dissipative nature the systems will be self-cooling, and a higher density area will form if there is no external energy input. Therefore, it is difficult to verify whether the binary collision assumptions are valid or not. In fact, for the inelastic collision process, under certain conditions, the assumptions of the presence of binary collisions translate to inelastic collapse of previously collided multiparticles.

References

[1] C. A. Coulomb, 'Sur une application des règles de Maximis et Minimis a quelques problèmes de stratique relatifs à l'Architecture', *Académie Royal Des Sciences Memoires de mathématique et de physique par divers savans*, 7, 343–382 (1773).

[2] T. Wakabayashi, 'Photoelastic method for determination of stress in powdered mass', in *Proceedings of the 9th Japan National Congress for Applied Mechanics*, 153–158 (1957).

[3] L. Vanel, D. Howell, D. Clark, R. P. Behringer and E. 'Clement, memories in sand: experimental tests of construction history on stress distributions under sandpiles', *Phys. Rev. E*, 60, R05040 (1999).

[4] P. W. Rowe, 'The stress-dilatancy relation for static equilibrium of an assembly of particles in contact', *Proc. R. Soc. A*, 269(1339), 500–527 (1962).

[5] R. A. Bagnold, 'Experiments on a gravity-free dispersion of large solid spheres in a Newtonian fluid under shear', *Proc. R. Soc. A*, 225, 49–63 (1954).

[6] Y. Forterre and O. Pouliquen, 'Flows of dense granular media', *Annu. Rev. Fluid Mech.*, 40, 1–24 (2008).

[7] G. De Josselin de Jong and A. Verruijt, 'Etude photo-élastique d'un empilement de disques', *Cahier Groupe Français Rhéologie*, 2, 73–86 (1969).

[8] A. Drecher and G. De Josselin de Jong, 'Photoelastic verification of a mechanical model for the flow of a granular material', *J. Mech. Phys. Solids*, 20, 337–351 (1972).

[9] M. Oda, J. Konishi and S. Nemat Nasser, 'Some experimentally based fundamental results on the mechanical behaviour of granular material: effects of particle rolling', *Geotechnique*, 30(4), 479–495 (1980).

[10] A. Drescher, 'An experimental investigation of flow rules for granular materials using optically sensitive glass particles', *Geotechnique*, 26(4), 591–901 (1976).

[11] H. G. B. Allersma, 'Optical analysis of stress and strain in photoelastic particle assemblies', PhD thesis, TUDelft, The Netherlands (1987).

[12] J.-P. Bouchaud, M. E. Cates and P. Claudin, 'Stress distribution in granular media and nonlinear wave equation', *J. Phys. (France) I*, 5, 639–656 (1995).

[13] P. Dantu, 'Contribution à l'étude mécanique et géométrique des milieux pulvérulents', in *Proceedings of the 4th International Conference on Soil Mechanics and Foundation Engineering*, Vol 1, Butterworth: London, 133 (1957).

[14] Y. Jiang and M. Liu, 'Granular solid hydrodynamics', *Granular Matter*, 11, 139–156 (2009).

[15] C. S. Campbell, 'Granular material flows: an overview', *Powder Technol.*, 162, 208–229 (2006).

[16] S. F. Edwards, R. B. S. Oakeshott, 'Theory of powders', *Phys. A*, 157, 1080–1090 (1989).

[17] M. Faraday, 'On a peculiar class of acoustical figures; and on certain forms assumed by groups of particles upon vibrating elastic surfaces', *Philos. Trans. R. Soc. London*, 52, 299–430 (1831).

[18] Y. Forterre and O. Pouliquen, 'Flows of dense granular media', *Annu. Rev. Fluid Mech.*, 40, 1–24 (2008).

[19] P. G. de Gennes, 'Granular matter: a tentative view', *Rev. Mod. Phys.*, 71(2), S374–S382 (1999).

[20] H. M. Jaeger, S. E. Nagel and R. P. Behringer, 'Granular solids, liquids, and gases', *Rev. Mod. Phys.*, 68, 1259–1271 (1996).

[21] O. Reynolds, 'On the dilatancy of media composed of rigid particles in contact', *Philos. Mag. Ser.*, 5(20), 469–481 (1885).

[22] S. Luding and H. J. Herrmann, 'Cluster growth in freely cooling granular media', *Chaos*, 9(3), 673–681 (1999).

Chapter 2

Contact mechanics of spherical particles

Granular material is defined as a collection of distinct macroscopic particles, for example, silt in soil and boulders in debris flow. The particle sizes can span 10^6 orders of magnitude, from micrometers to meters. The corresponding types of interactions between particles may change substantially. For example, in the micrometer order of magnitude, there are many observations that traditional contact mechanics cannot explain, such as the plastic yielding of contact surfaces under zero external loading, the jump-to-contact in the contact process, and the neck-separation in the separation process. All these phenomena are induced by the adhesion forces among surface atoms close to the contact area. Considering the meter order of magnitude, when compared with the elastic force, the surface force between particles is very weak and is usually disregarded. The traditional contact models focus primarily on nonadhesive contact where no tension force is allowed to occur within the contact area. Several analytical and numerical approaches have been proposed to solve contact problems in the absence of adhesion forces. The Hertzian contact model describes the relationship between the stress and the area of contact between two spheres of different radii by ignoring the adhesive surface forces, which provides a theoretical basis for the equations for load-bearing capabilities in bearings, gears, and any other bodies where two surfaces are in contact.

When two solid surfaces are brought into proximity to each other, they experience attractive van der Waals forces. The Bradley model provides a means of calculating the tensile force between two rigid spheres with perfectly smooth surfaces. The Johnson–Kendall–Roberts (JKR) model incorporates the effect of adhesion, existing only inside the area of contact, into the Hertzian contact model, by using a balance between the stored elastic energy and the loss in surface energy. The JKR model has successfully predicted the neck separation in the separation process arising from surface adhesion. The Derjaguin–Muller–Toporov (DMT) model assumes that the contact profile remains the same as in Hertzian contact but with additional attractive interactions outside the area of contact. When the particle surfaces separate, the DMT model simplifies to the Bradley model, that is, it ignores the deformation of particles caused by adhesion. The JKR model is applicable for particles with low elastic modulus, large size, and low adhesion force, while the DMT model is suitable for particles with high elastic modulus, small size, and high adhesion force.

The contact model for tangential forces between particles can become quite sophisticated. It is not only dependent on the normal force but is also impacted by its loading history. If surface adhesion is ignored, the tangential force can be described by the Mindlin–Deresiewicz (MD) model. If taking adhesion into account, the tangential force can be determined by the Thornton model, which combines the Savkoor–Brigg and MD models. All these contact models constitute a complete set of theory of the contact mechanics of spherical particles.

2.1 Nonadhesive contact

At each point of contact, the contact force can be resolved into normal and tangential components. It is assumed that the normal component is independent of the other load components, and this is evaluated by using the Hertz model. This assumption has been justified analytically for the case of identical particles, and numerically for dissimilar grain properties. Using a constitutive relation based on the MD model, we can compute the tangential force associated with frictional resistance from the relative lateral displacements and rotations of a pair of contacting particles.

2.1.1 Normal force (Hertz model)

Consider two rough solid spheres brought into physical contact through the action of applied forces. The contact between the two bodies occurs over many small areas, each of which constitutes a single asperity contact. When ignoring the surface adhesion, the contact force is generally calculated by the Hertz model if the following assumptions are satisfied: (1) the strains are sufficiently small for linear elasticity to be valid; (2) each particle can be considered as an elastic half-space, that is, the area of contact is much smaller than the characteristic radius of the body; (3) the surfaces are continuous and nonconforming; and (4) the surfaces are frictionless, so that only normal pressure is transmitted.

An elastic contact between two particles is shown in Figure 2.1. The relative approaches satisfy the following condition:

$$\alpha = R_1 + R_2 - |\vec{r}_1 - \vec{r}_2| > 0 \tag{2.1}$$

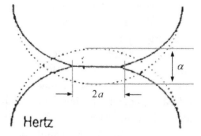

Figure 2.1: Schematic diagram of the elastic deformation of two spherical particles in the Hertz model. Dotted lines show the shapes of the undeformed spheres in the current configuration.

Where, R_1 and R_2 are the radii of the two spheres, and $\bar{r}_1$ and $\bar{r}_2$ the position vectors of the centers of the two spheres.

The contact area is circular and the contact radius is given as

$$a = \sqrt{\alpha R^*} \tag{2.2}$$

If surface adhesion is not considered, the normal contact force due to relative approach is, according to Hertz,

$$N = \frac{4}{3} E^* (R^*)^{1/2} \alpha^{3/2} \tag{2.3}$$

Substituting equations (2.2) to (2.3) then gives

$$N = \frac{4E^*}{3R^*} a^3 \tag{2.4}$$

Where, the effective Young's modulus E^* and the effective radius R^* are defined as

$$\frac{1}{R^*} \equiv \frac{1}{R_1} + \frac{1}{R_2} \tag{2.5}$$

$$\frac{1}{E^*} \equiv \frac{1 - \nu_1^2}{E_1} + \frac{1 - \nu_2^2}{E_2} \tag{2.6}$$

Where E_1, E_2, and ν_1; ν_2 are the Young's modulii and Poisson's ratios for two spheres, respectively.

If the increment of the relative approach is $\Delta\alpha$, it follows that the corresponding incremental normal force at the point of contact can be, according to equation (2.4), given as

$$\Delta N = 2\sqrt{\alpha R^*} E^* \Delta\alpha = 2aE^* \Delta\alpha \tag{2.7}$$

The above equations are also applicable to a situation of particle-boundary contact, by assigning an infinite radius to the planar wall.

If α is small enough, the normal force can be calculated in accordance with Hooke's law:

$$N \propto \alpha \tag{2.8}$$

This is a simplification of the more general nonlinear model of elasticity, such as the Hertz model. The fundamental assumptions of linear elasticity are the infinitesimal strains and linear relationships between the components of stress and strain. These assumptions are reasonable when constructing many engineering materials and in many engineering design scenarios. Linear elasticity is therefore used extensively in structural analysis and engineering design, often with the aid of finite element analysis.

2.1.2 Tangential force (Mindlin-Deresiewicz model)

In the case of zero adhesion, the tangential force at the contact of two spheres is modeled by the MD model, which predicts that if two contacting surfaces are subjected to an increasing tangential displacement δ, the relative sliding initiates at the perimeter and progresses inward over an annular area of the contact surface (Figure 2.2). The incremental tangential force ΔT, due to the incremental tangential displacement $\Delta\delta$, depends not only on the loading history but also on the variation in the normal force. In all cases, the incremental tangential force can be obtained from the following equation:

$$\Delta T = 8aG^*\theta_k\Delta\delta + (-1)^k\,\mu(1-\theta_k)\Delta N \tag{2.9}$$

Where $k = 0$, 1, and 2 denotes the paths of loading, unloading, and reloading, respectively. If $|\Delta T| < \mu\Delta N$, then

$$\theta_k = 1 \tag{2.10}$$

If $|\Delta T| < \mu\Delta N$, then

$$\theta_k = \begin{cases} \left(1 - \frac{T + \mu\Delta N}{\mu N}\right)^{1/3}, & k = 0 \\[2mm] \left(1 - \frac{(-1)^k(T-T_k)+2\mu\Delta N}{2\mu N}\right)^{1/3}, & k = 1, 2 \end{cases} \tag{2.11}$$

Where μ is the friction coefficient, N is the normal force, T_k represents the historical tangential forces from which loading or reloading commenced, and the effective shear modulus G^* is defined as:

$$G^* \equiv \frac{2 - \nu_1}{G_1} + \frac{2 - \nu_2}{G_2} \tag{2.12}$$

Here, G_1 and G_2 are the shear modulii for the two spheres. It can be determined by:

$$G_1 \equiv \frac{E_1}{2(1 + \nu_1)} \tag{2.13}$$

$$G_2 \equiv \frac{E_2}{2(1 + \nu_2)} \tag{2.14}$$

It can be seen from equation (2.9) that the tangential force increment ΔT is not only dependent on the normal force, but is also related to the loading history of the tangential force. The normal force and tangential force are coupled together. In addition, the contact force is usually a nonlinear function of the deformation. It is difficult to obtain the analytical solutions of normal and tangential components of the contact force. Therefore, to simplify the solution procedure, an incremental method is generally adopted to calculate the deformation, the normal force, and tangential force.

Both the Hertz and MD models have been developed for purely elastic contacts. Indeed, there occurs a plastic deformation essentially during the contact process. At the loading stage, the normal force increases approximately linearly with the relative approach, indicating a nearly constant effective Young's modulus. However, the unloading follows a different curve, again nearly linear, but with a steeper slope that indicates a larger but still constant effective Young's modulus. It is observed that a small flat indentation remains around the contact area, therefore, the contact is quickly lost and the contact force reduces to zero faster, as shown in Figure 2.3. In addition, when successive contact situations are imposed, the particle surface becomes hard. It is also possible that with a gradually increasing loading of the tangential force, until the two particle surfaces come into contact, a relative sliding motion occurs that erases tiny bumps on their surfaces. Therefore, each

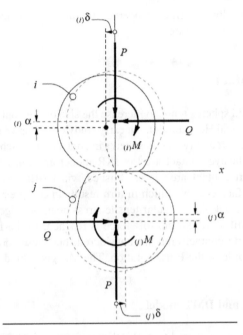

Figure 2.2: Schematic diagram of tangential contact deformation of two spherical particles subjected to normal and tangential forces. Dotted lines show the shapes of the undeformed spheres in the current configuration.

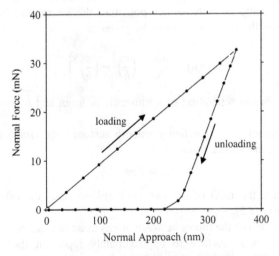

Figure 2.3: Relationship between contact force and relative approach of elasto-plastic particles during loading and unloading.

contact can change the local surface properties in such a way that these properties may change with both position and time.

2.2 Adhesive contact

Adhesion between solid spheres tends to increase the size of the contact zone beyond what is predicted by the traditional Hertz model. An early study on adhesive contact was the Bradley model for the adhesive force of two rigid spheres in contact with each other. The most widely known models for adhesive contact are the JKR model and the DMT model. In the JKR model, an equilibrium contact area is established, via Griffith energy balance, between elastic energy and surface energy, which in turn results in compressive stress in the central region of contact and crack-like singular tensile stress near the edge of contact; the contact area remains finite until a critical pull-off force is reached. In the DMT model, molecular forces outside the Hertz contact area are considered, but these forces are assumed to not change the contact profile of the Hertz solution. The Maugis–Dugdale model links the JKR and DMT models.

2.2.1 Bradley model and DMT model

When two solid bodies are adhered to each other to form an interface, the Dupré energy of adhesion is

$$\Delta\gamma \equiv \gamma_1 + \gamma_2 - \gamma_{12} \tag{2.15}$$

In which γ_1 and γ_2 denote free energy of the two solid surfaces, and γ_{12} is the interfacial energy. According to the Lennard–Jones potential, the interaction per unit area with a distance of h between the two surfaces can be given as:

$$\sigma(h) = \frac{8\Delta\gamma}{3\varepsilon}\left[\left(\frac{\varepsilon}{h}\right)^3 - \left(\frac{\varepsilon}{h}\right)^9\right] \tag{2.16}$$

Here, ε is the equilibrium separation of a molecule or atom and h the separation between atomic planes.

For two rigid particles with perfectly smooth surfaces, the Bradley model provides a means of calculating the tensile force

$$F_0 = 2\pi R^* \Delta\gamma \tag{2.17}$$

Where R_1 and R_2 are the radii of the two rigid spheres, and the effective radius is R^*, according to equation (2.5).

By regarding the sum of the forces between atoms as an interaction between the surfaces, and by using a parabolic surface to approximately represent the sphere surface, the interaction between the two spherical particles at $h = r^2/2R$ is:

$$F = \int_0^\infty 2\pi\sigma(h)dr = 2\pi R \int_0^\infty \sigma(h)dh \tag{2.18}$$

Substituting equation (2.15) into equation (2.18), and assuming the gap between the two sphere surfaces is h_0, the adhesion force for rigid spheres F is derived as:

$$F = \frac{2}{3}\left[4\left(\frac{\varepsilon}{h_0}\right)^2 - \left(\frac{\varepsilon}{h_0}\right)^8\right]\pi R^* \Delta\gamma \qquad (2.19)$$

Equation (2.19) is the so-called Bradley model. When $h_0 = \varepsilon$, equation (2.17) can be obtained.

Experimental measurements show that when external loading is smaller, the contact area of the particles is greater than the one determined by the Hertz model, and when the loading gradually decreases to zero, the contact area tends to remain at a constant value. This indicates that there exists a surface attraction between the solid bodies; this interaction becomes crucial when the loading is reduced to zero. The DMT model assumes that the contact profile remains the same as in Hertzian contact, but with additional attractive interactions outside the area of contact. On the basis of equation (2.17), the DMT model incorporates adhesion into Hertzian contact, as shown in Figure 2.4, which obtains the relationship between the radius of the contact surface.

$$a^3 = \frac{3R^*}{4E^*}\left(N + 2\pi R^* \Delta\gamma\right) \qquad (2.20)$$

It can be seen from equation 2.20 that the radius of the contact zone caused by attraction is:

$$a^3 = \frac{3\pi\left(R^*\right)^2 \Delta\gamma}{2E^*} \qquad (2.21)$$

In addition, the minimal loading N can be obtained from equation (2.20), which corresponds to the pull-off force N_c that separates the adhesive surfaces.

$$N_c = 2\pi R^* \Delta\gamma \qquad (2.22)$$

The DMT model takes into account the additional attractive interactions outside the area of contact. When the particle surfaces separate, the DMT model is simplified to the Bradley model. The Bradley model regards a particle as rigid body, that is, it ignores the deformation of particles caused by adhesion.

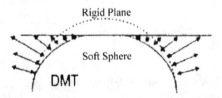

Figure 2.4: Schematic diagram of contact deformation of spherical particles with a flat wall in the DMT model. There exists attraction outside the area of contact. Dotted lines show the shapes of the undeformed spheres in the current configuration.

2.2.2 JKR model

In contrast to the DMT model, the JKR model assumes that the adhesive forces are confined to inside the area of contact, as shown in Figure 2.5. If there is no adhesion between particles, the radius a_0 of the two-particle contact surface is given by the Hertz model; if there is adhesion between particles, despite the external loading being N, the radius of contact surface $a > a_0$.

Assuming the contact radius is a, the surface energy, U_s equals $-\pi a^2 \Delta \gamma$. The total energy U_T is the function of the contact area A. The balance between the stored elastic energy and the loss in surface energy is reached when $dU_T/dA = 0$, and then the JKR model can be formulated. Under the combined action of the external loading N, and surface adhesion force, according to the JKR model, the effective normal force N_1 is defined as:

$$N_1 = N + 3\pi R^* \Delta \gamma + \sqrt{\left(3\pi R^* \Delta \gamma\right)^2 + 6\pi R^* \Delta \gamma N} \tag{2.23}$$

The Hertz model for the radius of contact between two spheres, modified to take into account the surface energy, has the form:

$$a^3 = \frac{3R^*}{4E^*}\left(N + 3\pi R^* \Delta \gamma + \sqrt{\left(3\pi R^* \Delta \gamma\right)^2 + 6\pi R^* \Delta \gamma N}\right) \tag{2.24}$$

The relative approach α of the two particles is:

$$\alpha = \frac{a^2}{R^*} - \left(\frac{2\pi a \Delta \gamma}{E^*}\right)^{1/2} \tag{2.25}$$

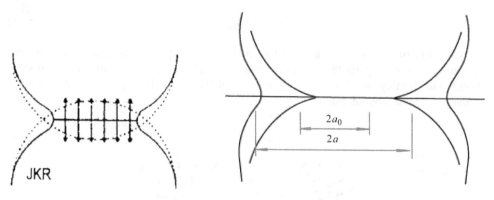

Figure 2.5: Schematic diagram of contact deformation of spherical particles with a flat wall in the JKR model. (a) There exists attraction inside the area of contact. Dotted lines show the shapes of the undeformed spheres in the current configuration. (b) Variation of contact radius under the same external loading; a is the radius with adhesion force, a_0 is the radius without adhesion force.

If the increment of the relative approach between the two contacting spheres is $\Delta\alpha$, it follows that the corresponding incremental normal force at the contact can be given as:

$$\Delta N = 2aE^*\Delta\alpha\left(\frac{3\sqrt{N} - 3\sqrt{N_c}}{3\sqrt{N} - \sqrt{N_c}}\right) \tag{2.26}$$

By analyzing equation (2.24), the following conclusions may be drawn:

1. When the interfacial energy is zero, $\gamma = 0$, the Hertz equation for contact between two spheres is recovered. The contact radius is

$$a^3 = \frac{3R^*N}{4E^*}$$

2. When the applied load $N = 0$, the contact radius is:

$$a^3 = \frac{9\pi\Delta\gamma(R^*)^2}{2E^*} \tag{2.27}$$

3. When $(3\pi R^*\Delta\gamma)^2 + 6\pi R^*\Delta\gamma N \geq 0$, equation (2.24) has a solution, that is, N satisfies:

$$N \geq -3\pi R^*\Delta\gamma/2 \tag{2.28}$$

Its physical meaning is that when the external loading N is negative (i.e., separation of the two particles in contact), the radius of the contact surface decreases. If N increases until $N = -3\pi R^*\Delta\gamma/2$, the adhesion between particles reaches a critical state, that is to say, if N becomes larger, then the two particles are separated. Thus, the maximal N to separate the particles, that is, N_c, is given by:

$$N_c = \frac{3\pi R^*\Delta\gamma}{2} \tag{2.29}$$

The corresponding contact radius is:

$$a_c^3 = \frac{3R^*N_c}{4E^*} = \frac{9\pi(R^*)^2\Delta\gamma}{8E^*} \tag{2.30}$$

The relationship between N_c and a_c is

$$\left(\frac{N}{N_c} - \frac{a^3}{a_c^3}\right)^2 = 4\left(\frac{a}{a_c}\right)^3 \tag{2.31}$$

The variation of the ratio a/a_c with N/N_c is shown in Figure 2.6.

It can be seen from Figure 2.6 that when N is decreased up to zero, the contact radius decreases. The particle surface is still adhered as indicated by point A.

$$\left.\left(\frac{a}{a_c}\right)\right|_{N=0} = \sqrt[3]{4} \approx 1.587 \tag{2.32}$$

When N becomes negative, the contact surface area continues to decrease. At point B the pulling force reaches a peak with $N = -N_c$, $a = a_c$. At this stage, the contact between

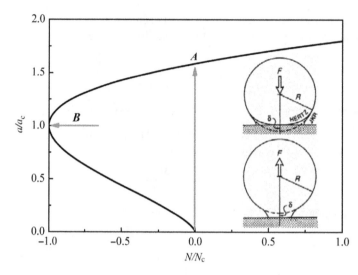

Figure 2.6: The variation of the radius of the contact surface with the external loading according to the JKR model.

particles begins to lose stability, although the surfaces are still adhered together. If N_c is maintained, the two particles are certainly separated immediately. But, if N gradually decreases at this time, the radius of the contact surface decreases gradually and stably until complete separation.

It can be seen from equations (2.22) and (2.29) that the pulling off force N_c, obtained by the DMT model and the JKR model, is not dependent on elastic modulus E, which makes these models suitable for rigid ball contact. The problem here is that the pre-factors in the two models are inconsistent.

2.2.3 Maugis–Dugdale model

The apparent contradiction between the DMT model and the JKR model is well explained after the introduction of the Tabor coefficient (μ) defined as:

$$\mu \equiv \left(\frac{R^* \Delta \gamma^2}{E^{*2} \varepsilon^3} \right)^{1/3} \tag{2.33}$$

Here, ε is the equilibrium separation between the two surfaces in contact.

The Tabor coefficient can be regarded as the ratio of the effective ranges of the elastic deformation force and the surface force caused by adhesion. When $\mu < 0.1$, namely, the particles exist with a small radius of curvature, a high adhesion, and a high elastic modulus, this is suitable for the DMT model; when $\mu > 5$, namely, the particles exist with a large radius, a low adhesion, and a low elastic modulus, this is suitable for the JKR model. Thus, according to the Tabor coefficient, the two models are the two extreme limits.

A dimensionless parameter that is equivalent to the Tabor coefficient is defined as:

$$\lambda \equiv \sigma_0 \left(\frac{9R^*}{2\pi\Delta\gamma E^{*2}} \right)^{1/3} \tag{2.34}$$

Where $\sigma_0 = \frac{16}{9\sqrt{3}} \frac{\Delta\gamma}{\varepsilon}$, then

$$\lambda = 1.16\mu \tag{2.35}$$

In the Maugis–Dugdale model, the surface traction distribution is divided into two parts: one due to the Hertz contact pressure, and the other due to the Dugdale adhesive stress, as shown in Figure 2.7. When $r < a$, the round area corresponds to the actual contact surface under the combined effects of an external force and a surface force; when $a < r < c$, the annulus area corresponds to the separation distance of the two surfaces from zero to 0.971ε.

The relationships of λ, loading, and relative approach with the radius of the contact surface can be obtained through the Maugis–Dugdale model:

$$\frac{1}{2}\lambda\bar{a}^2 \left[(m^2 - 2)\arccos(1/m) + \sqrt{m^2 - 1} \right] + \frac{4}{3}\lambda^2\bar{a} \left[(m^2 - 2)\arccos(1/m) - m + 1 \right] = 1 \tag{2.36}$$

$$\bar{N} = \bar{a}^3 - \lambda\bar{a}^2 \left[\sqrt{m^2 - 1} + m^2 \arccos(1/m) \right] \tag{2.37}$$

$$\bar{\alpha} = \bar{a}^2 - \frac{4}{3}\bar{a}\lambda\sqrt{m^2 - 1} \tag{2.38}$$

In these formulae:

$$\bar{a} = a \left(4E^*/3\pi\Delta\gamma R^2 \right)^{1/3} \tag{2.39}$$

$$\bar{N} = N/\pi R\Delta\gamma \tag{2.40}$$

$$\bar{\alpha} = \alpha \left(16E^{*2}/9\pi^2\Delta\gamma^2 R \right)^{1/3} \tag{2.41}$$

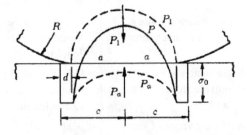

Figure 2.7: Schematic of contact area for the Maugis–Dugdale model, in which P_1 is Hertz tension applied on the area with the radius of a; P_a is the adhesion force imposed on the area with the radius of c.

$$\alpha = \frac{a^2}{R} - 2\frac{\sigma_0}{E^*}\sqrt{c^2 - a^2} \tag{2.42}$$

$$m = c/a \tag{2.43}$$

$$c = a + 0.971\varepsilon \tag{2.44}$$

From the mechanism of surface attraction, the JKR model considers only the adhesion effect inside the contact surface, and the DMT model considers only the adhesion effect outside the contact surface, whereas the Maugis–Dugdale model describes the adhesion effect using a square-well potential. The assumptions and limitations of these contact models are listed in Table 2.1.

The Maugis–Dugdale model is more general, which can be used to describe contact problems concerning nearly all materials. The JKR model and the DMT model are the upper and lower limits of the Maugis–Dugdale model, respectively (Table 2.2). However, when the Maugis–Dugdale model has no analytical solutions, the approximate solution is usually adopted; see Figure 2.8.

It can be seen from Figure 2.8 that for adhesive contact, the radius of contact surface is greater than that predicted by the Hertz model. For a given loading, the contact radius increases with the decrease in λ, which indicates that the adhesion effect is more marked. In other words, even when the loading drops to zero or negative, a certain contact surface continues to exist.

It is shown in Figure 2.9 that under high loading, such as, $\frac{N}{\pi R^* \Delta \gamma} \approx 40 \sim 2500$, the Hertz model is appropriate. The Maugis–Dugdale area is located between the DMT and the JKR areas. If the adhesion deformation is less than 0.05ε, the Maugis–Dugdale area transits to the DMT area. If the adhesion deformation is greater than 20ε, the Maugis–Dugdale area transits to the JKR area. These three models are applicable to situations in which the elastic

Table 2.1: Comparison of some contact models

Contact model	Major assumption	Limitation
Hertz	Ignores surface force	It is not applicable when there is a surface force and the loading is low.
JKR	Short-distance force on the contact force	It is applicable in situations with greater value of λ, but underestimates the loading magnitude.
DMT	Long-distance force outside the contact surface	It is applicable in situations with smaller value of λ, which under estimates the contact area.
Maugis–Dugdale	Surface energy of the contact face is described by a square-well potential	It is applicable for λ under any circumstances, but the equation has many parameters and the analytical solution is hard to obtain.

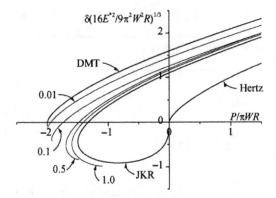

Figure 2.8: Comparison of the relative approach and the normal force calculated by the Hertz model, the JKR model, the DMT model, and the Maugis–Dugdale model; λ is 0.01, 0.1, 0.5, and 1.0, respectively. The DMT model and the JKR model are the two limiting cases of the Maugis–Dugdale model.

contact and relative approach are much smaller than the particle size and the tangential motion is negligible.

2.2.4 Thornton model

The JKR model extends the Hertz theory to include attractive surface forces but does not account for shear stress. The Savkoor–Briggs model extends the JKR theory by accounting for the presence of shear stress at the adhesive interface. In this analysis, the shear force initiates a reduction of the radius of contact as predicted by JKR theory, and thus a reduction of the adhesive force in the normal direction. By combining the Savkoor–Briggs model and the MD model, the Thornton model comprehensively takes into account the influences of

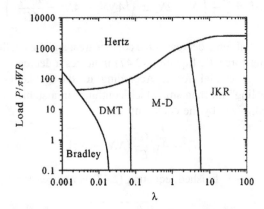

Figure 2.9: Scope of application of the Hertz model, the JKR model, the DMT model, and the Maugis–Dugdale model.

Table 2.2: Summary of Hertz model, JKR model, DMT model, and Maugis–Dugdale model

Contact model	λ	Radius of contact surface	Elastic relative approach
Hertz	–	$a^3 = \dfrac{3R^*}{4E^*}N$	$\alpha = \dfrac{a^2}{R}$
DMT	$\lambda < 0.1$	$a^3 = \dfrac{3R^*}{4E^*}(N + 2\pi\Delta\gamma R)$	$\alpha = \dfrac{a^2}{R}$
JKR	$\lambda > 5$	$a^3 = \dfrac{3R^*}{4E^*}\left(N + 3\pi\Delta\gamma R + \sqrt{6\pi\Delta\gamma RN + 9\pi^2\Delta\gamma^2 R^2}\right)$	$\alpha = \dfrac{a^2}{R} - \sqrt{\dfrac{2\pi\Delta\gamma a}{E^*}}$

particle plastic deformation and loading history, and constructs a set of equations for the tangential force.

2.2.4.1 From contact initiation to a critical peeling state

At an initial "peeling" phase of contact, a decrease in the contact area occurs with increasing load T, without evidence of slip. Assuming the tangential displacement is δ and the increment is $\Delta\delta$, (the tangential force and its increment are given by:

$$T = 8G^*a\delta \tag{2.45}$$

$$\Delta T = 8G^*a\Delta\delta \tag{2.46}$$

It was also suggested that the application of a tangential force reduces the potential energy by an amount $T\delta/2$, leading to the following expression for the radius of the contact area:

$$a^3 = \frac{3R^*}{4E^*}\left(N + 2N_c \pm \sqrt{4NN_c + 4N_c^2 - \frac{T^2E^*}{4G^*}}\right) \tag{2.47}$$

The peeling tangential force decreases the contact area, and the peeling process continues up to a critical peeling force T_c. Equation (2.47) indicates a decrease in the contact radius under an increasing tangential force. According to the Savkoor–Briggs model, this corresponds to a "peeling" mechanism, which continues in a stable manner until a critical value of T_c is reached, given by the equation:

$$T_c = 4\sqrt{\frac{G^*}{E^*}\left(NN_c + N_c^2\right)} \tag{2.48}$$

When the peeling process is finished, equation (2.47) is reduced to:

$$a_p = \sqrt[3]{\frac{3R^*}{4E^*}(N + 2N_c)} \tag{2.49}$$

2.2.4.2 After the critical peeling state

If the relative tangential displacement increases beyond that corresponding to $T = T_c$, the peeling process is complete. The rate of energy release is more rapid than the rate at which the work of adhesion can absorb the energy, and hence, the quasi-static approach becomes inappropriate. This is because kinetic energy terms begin to play a significant role. It is argued that it is reasonable to expect that the contact area would diminish to the Hertzian area when $T = T_c$. It is also suggested that, for larger values of $T > T_c$, a shear mode of separation may occur, resulting in a behavior somewhat similar to Mindlin's concept of slip. According to the Thornton model, the tangential force between particles can be determined by the ratio of the critical peeling force and sliding friction, and then the effective normal force is computed as:

$$N_2 = N_1 \left[1 - (N_1 - N)/3N_1 \right]^{3/2} \tag{2.50}$$

In which $N_1 = N_0 + 2N_c + \sqrt{4N_c^2 + 4NN_c^2}$ is the equivalent loading force, see equation (2.23). $N_c = 3\pi R^* \Delta\gamma/2$ is the pulling-off force; see equation (2.29).

The sliding friction force is:

$$T_s = \mu N_2 = \mu N_1 \left[1 - (N_1 - N)/3N_1 \right]^{3/2} \tag{2.51}$$

If $T_c > T_s$, the tangential force is equal to the sliding friction force (Figure 2.10):

$$T = \mu N_1 \left[1 - (N_1 - N)/3N_1 \right]^{3/2} \tag{2.52}$$

If $T_c < T_s$, the sliding circle expands sharply inward, T increases gradually to T_s, and the sliding reoccurs. This process can be calculated by the partial sliding method in the MD model. When the peeling process ends, if there is a relative tangential displacement increment $\Delta\delta$ between the two contact surfaces, the corresponding tangential force

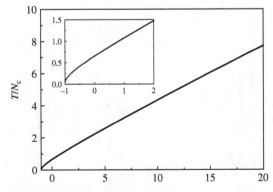

Figure 2.10: When $\mu = 0.3$, variation of the normal external loading with tangential force is obtained by equation (2.52).

Table 2.3: Summary of contact models of spherical particles

Contact force	Without adhesion	With adhesion	Remarks
Normal component	By the Hertz model For normal relative approach α, the contact force is: $$N = \frac{4}{3}E^*(R^*)^{1/2}\alpha^{3/2}$$ For the relative approach increment $\Delta\alpha$, the incremental normal force is: $$\Delta N = 2\alpha E^* \Delta\alpha$$	By the JKR model For relative approach α, the contact force N is: $$\alpha = \frac{a^2}{R^*} - \left(\frac{2\pi a \Delta\gamma}{E^*}\right)^{1/2}$$ $$a^3 = \frac{3R^*}{4E^*}\left(N + 3\pi R^*\Delta\gamma + \sqrt{(3\pi R^*\Delta\gamma)^2 + 6\pi R^*\Delta\gamma N}\right)$$ For the relative approach increment $\Delta\alpha$, the normal force increment $$\Delta N = 2aE^*\Delta\alpha\left(\frac{3\sqrt{N_1}-3\sqrt{N_c}}{3\sqrt{N_1}-\sqrt{N_c}}\right)$$ where $N_1 = N_0 + 2N_c + \sqrt{4N_c^2 + 4NN_c}$, $N_c = 3\pi R^* \Gamma/2$	1. Relative approach $\alpha = R_1 + R_2 - \|\bar{r}_1 - \bar{r}_2\|$ R_1, R_2 and $\bar{r}_1, \bar{r}_2$ are the radii and center vectors of the two particles, respectively. 2. Radius of the contact surface $a = \sqrt{\alpha R^*}$ 3. R^* and E^* are the effective radius and effective Young's modulus of the particles. $$\frac{1}{R^*} = \frac{1}{R_1} + \frac{1}{R_2}$$ $$\frac{1}{E^*} = \frac{1-\nu_1^2}{E_1} + \frac{1-\nu_2^2}{E_2}$$ Where E_1, ν_1, and E_2, ν_2 are Young's moduli and Poisson's ratios, respectively, of particles 1 and 2. 4. G^* is the effective shear modulus $$G^* = \frac{2-\nu_1}{G_1} + \frac{2-\nu_2}{G_2}$$ G_1 and G_2 are the shear moduli of the two particles. The relationships of shear moduli G_1 and G_2 with Young's modulus and Poisson's ratio are: $G_1 = E_1/2(1+\nu_1)$; $G_2 = E_2/2(1+\nu_2)$
Tangential component	By the Mindlin–Deresiewicz model If the tangential displacement increment is $\delta\Delta$, the tangential force increment is: $$\Delta T = 8aG^*\theta_k\Delta\delta + (-1)^k\mu(1-\theta_k)\Delta N$$ $k = 0, 1,$ and 2 denotes the path of loading, unloading, and reloading, respectively. If $\|\Delta T\| < \mu\Delta N$, $\theta_k = 1$ If $\|\Delta T\| \geq \mu\Delta N$, $$\theta_k = \begin{cases} \left(1 - \dfrac{T + \mu\Delta N}{\mu N}\right)^{1/3}, & k = 0 \\ \left(1 - \dfrac{(-1)^k(T-T_k)+2\mu\Delta N}{2\mu N}\right)^{1/3}, & k = 1, 2 \end{cases}$$ where μ is the friction coefficient, and N the normal component of external loading.	By the Thornton model 1. From contact initiation to critical peeling state When the tangential displacement is δ and the increment is $\Delta\delta$, the tangential force and its increment are: $$T = 8G^*a\delta, \quad \Delta T = 8G^*a\Delta\delta$$ The radius of the contact force $a^3 = (3R^*/4E^*)\times$ $$\left(N + 2N_c \pm \sqrt{4NN_c + 4N_c^2 - (T^2E^*/4G)}\right)$$ Critical peeling force $T_c = 4\sqrt{G^*(NN_c + N_c^2)}/E^*$ 2.1 After critical peeling, the sliding friction force is: $$T_s = \mu N_2 = \mu N_1[1 - (N_1 - N)/3N_1]^{3/2}$$ In which N_1 and N_c are defined in the nonadhesive part. If $T_c \geq T_s$, the tangential force is: $$T = \mu N_1[1 - (N_1 - N)/3N_1]^{3/2}$$ If $T_c < T_s$, the tangential force increment can be calculated by the equation for nonadhesive contact (note that N is changed to $(N + 2N_c)$; when T gradually increases to $T \geq T_s$, the slip occurs.	

increment ΔT is:

$$\Delta T = 8aG^*\theta_k\Delta\delta + (-1)^k\mu(1 - \theta_k)\Delta N \qquad (2.53)$$

Where $k = 0$, 1, and 2 denotes the path of loading, unloading, and reloading, respectively. If $|\Delta T| < \mu\Delta N$, then

$$\theta_k = 1 \qquad (2.54)$$

If $|\Delta T| \geq \mu\Delta N$, then

$$\theta_k = \begin{cases} \left(1 - \dfrac{T + \mu\Delta N}{\mu(N + N_c)}\right)^{1/3}, & k = 0 \\[4mm] \left(1 - \dfrac{(-1)^k(T - T_k) + 2\mu\Delta N}{2\mu(N + N_c)}\right)^{1/3}, & k = 1, 2 \end{cases} \qquad (2.55)$$

In which μ is the friction coefficient, and N the normal component of external loading.

When T_k increases to T_s, there is a slip between the two particles, and the tangential force becomes the sliding friction force.

$$T = \mu N_1 \left[1 - (N_1 - N)/3N_1\right]^{3/2} \qquad (2.56)$$

References

[1] B. V. Derjaguin, V. M. Muller and Y. P. Toporov, 'Effect of contact deformations on the adhesion of particles', *J. Colloid Interface Sci.*, 53(2), 314–326 (1975).

[2] K. L. Johnson, *Contact Mechanics*, Cambridge University Press, Cambridge, UK (1985).

[3] K. L. Johnson, K. Kendall and A. D. Roberts, 'Surface energy and the contact of elastic solids', *Proc. R. Soc. Lond. A*, 324, 301–313 (1971).

[4] L. D. Landau and E. M. Lifshitz, *Theory of Elasticity*, Pergamon Press, Oxford, UK (1986).

[5] D. Maugis, 'Adhesion of spheres: the JKR-DMT transition using a dugdale model', *J. Colloid Interface Sci.*, 150, 243–269 (1992).

[6] R. D. Mindlin and H. Deresiewicz, 'Elastic spheres in contact under varying oblique forces', *J. Appl. Mech.*, 20, 327–344 (1953).

[7] A. R. Savkoor and G. A. D. Briggs, 'The effect of tangential force on the contact of elastic solids in adhesion,' *Proc. R. Soc. Lond. A*, 356, 103–114 (1977).

[8] C. Thornton, 'Interparticle sliding in the presence of adhesion', *J. Phys. D: Appl. Phys.*, 24, 1942–1946 (1991).

[9] D. Tabor, 'Surface forces and surface interactions', *J. Colloid Interface Sci.*, 58, 2–13 (1977).

[10] W. N. Unertl, 'Implications of contact mechanics models for mechanical properties measurements using scanning force microscopy', *J. Vac. Sci. Technol. A*, 17(4), 1779–1786 (1999).

Chapter 3

Soft-sphere approach and hard-sphere approach

The accurately computation of inter-particle forces requires sophisticated procedures. Therefore, several simplified approaches, that do not introduce significant errors, have been proposed. On the basis of the mechanism of contact forces, a granular system may be modeled as either hard spheres or as soft spheres, which are the two frequently used types of approximate formulations for modeling the inter-particle forces. In both of these approaches, no effort is made to account for the sphere deformation resulting from contact or collision. However, this level of detail has been found to be unnecessary to understand the macroscopic behavior of granular assemblies.

The soft-sphere approach uses a linear-spring/dashpot scheme, and introduces parameters such as the spring stiffness and the damping coefficient. This approach ignores the particle surface deformation and the loading history, and it calculates the contact force based on the relative approach between particles and the tangential displacement. The hard-sphere approach completely ignores the details of contact force evolution, and thus the contact process is simplified to an instantaneous collision process. The after-collision speed is directly generated. A restitution coefficient is introduced to represent the energy dissipation during the collision process.

3.1 Soft-sphere approach

The soft-sphere approach was originally developed by Cundall and Strack in 1978 [3]. In soft-sphere approaches, the particles are allowed to approach relative to each other, and using a contact force scheme, the contact forces are subsequently calculated from the deformation history of the contact. The soft-sphere approaches allow for the relative approach of multiple particles, although the net contact force is obtained from the addition of all pair-wise interactions. The soft-sphere approaches are essentially time-driven, where the time step should be carefully chosen when calculating the contact force. During the past decade, hundreds of research papers have been published on the application of soft-sphere approaches toward a variety of problems (from geomechanics to chemical engineering). The soft-sphere approaches are capable of handling multiple particle contacts and of resolving the resultant equilibrium condition. This is important for handling dense and quasi-static systems. This approach may be best suited for dense granular assemblies with large deformations in unusual geometrical configurations.

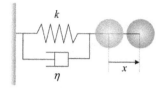

Figure 3.1: The soft-sphere approach simplifies the contact process between particles as a damped spring system.

As shown in Figure 3.1, a linear-spring/dashpot model is employed to calculate the contact forces. The translational motion of particles, for example in x direction, is given by

$$m\ddot{x} + \eta\dot{x} + kx = 0 \qquad (3.1)$$

where k is the spring stiffness, and η is the damping coefficient.

It can be seen from equation (3.1) that the restoring force is proportional to the displacement of the particle, and the viscous resistance is proportional to the speed, but in the opposite direction. Therefore, there is a gradual decay in the system energy. With a progressive increase in damping, the system shows under-damping vibration, critical damping vibration, and over-damping vibration, respectively.

As shown in Figure 3.2, particle i contacts particle j at point C. The dotted line indicates the surface of particle i when the contact begins. With the relative motion of the two particles, the particle surface is gradually deformed and generates a contact force. Since the soft-sphere approach does not consider the details of the deformation, the normal and tangential components of the contact force are expressed directly in terms of the normal relative approach and the tangential relative approach.

Soft-sphere approaches set the spring, damper, slider, coupler, and other parameters between particle i and particle j. The coupler is used to determine the pair correlation between the contacted particles, without leading to any force. As regards to the tangential component, if the tangential force is more than the yield value, then two particles slip under the normal force and friction; the sliding resistance device is used to achieve this purpose.

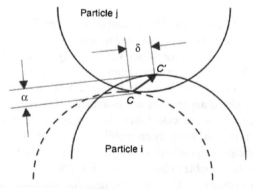

Figure 3.2: Soft-sphere approach considers the normal relative approach α and tangential displacement δ of the two particles while ignoring the surface deformation.

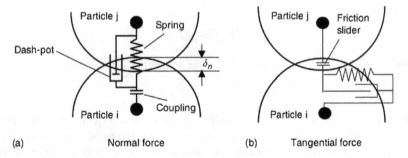

(a)   Normal force   (b)   Tangential force

Figure 3.3: Soft-sphere approach simplifies the contact force between particles.

The soft-sphere approach requires the introduction of spring stiffness k and damping coefficient η, in order to quantify the function of springs, dampers, and sliders, as shown in Figure 3.3.

3.1.1 Calculation of contact force

The normal force $\vec{F}_{nij}$ is the resultant force of the elastic force and damping force imposed on the particle i by the spring and damper, respectively, as shown in Figures 3.2 and 3.3. For two-dimensional particles, the elastic force is proportional to the relative approach, and the damping force is proportional to the relative velocity of particles. Therefore, $\vec{F}_{nij}$ is expressed as:

$$\vec{F}_{nij} = \left(-k_n\alpha - \eta_{ni}\vec{G}\vec{n}\right)\vec{n} \tag{3.2}$$

where α is the normal relative approach, $\vec{G}$ is the speed of particles i relative to particle j; $\vec{G} = \vec{v}_i - \vec{v}_j$. $\vec{n}$ is the unit vector from the center of particle i to the one of particle j. k_n and η_{ni} are, respectively, the normal spring stiffness and damping coefficient of particle i.

For the three-dimensional particles, according to the Hertz model, $\vec{F}_{nij}$ is expressed as:

$$\vec{F}_{nij} = \left(-k_n\alpha^{3/2} - \eta_{ni}\vec{G}\vec{n}\right)\vec{n} \tag{3.3}$$

Similar to equation (3.2), the tangential force $\vec{F}_{tij}$ is given by:

$$\vec{F}_{tij} = -k_t\vec{\delta} - \eta_{tj}\vec{G}_{ct} \tag{3.4}$$

where k_t and η_{tj} are the tangential elastic and damping coefficient, respectively; $\vec{G}_{ct}$ is the slip velocity of the contact point; and $\vec{\delta}$ is the tangential displacement of the contact point, which is not necessarily maintained in the same direction as the slip velocity $\vec{G}_{ct}$ in a three-dimensional movement.

$$\vec{G}_{ct} = \vec{G} - \left(\vec{G}\vec{n}\right)\vec{n} + a_i\vec{\Omega}_i \times \vec{n} + a_j\vec{\Omega}_j \times \vec{n} \tag{3.5}$$

where a_i and a_j are the radii of the particle i and particle j; $\vec{\Omega}_i$ and $\vec{\Omega}_j$ are the angular velocities of particle i and particle j.

If the following relation is satisfied:

$$\left|\vec{F}_{tij}\right| > \mu\left|\vec{F}_{nij}\right| \tag{3.6}$$

then particle i slips relative to particle j, and the tangential force becomes:

$$\vec{F}_{tij} = -\mu\left|\vec{F}_{nij}\right|\vec{t} \tag{3.7}$$

The above equation is actually the Coulomb's law of friction. Friction is not a fundamental force, but instead occurs because of the electromagnetic forces between charged particles that constitute the surfaces in contact. Because of the complexity of these interactions, friction cannot be calculated from first principles. Instead, the coefficient of friction μ must be found empirically.

The tangential unit vector $\vec{t}$ is determined by:

$$\vec{t} = \frac{\vec{G}_{ct}}{\left|\vec{G}_{ct}\right|} \tag{3.8}$$

The resultant force and moment on particle i are:

$$\vec{F}_{ij} = \vec{F}_{nij} + \vec{F}_{tij}; \quad \vec{T}_{ij} = a_i\vec{n} \times \vec{F}_{tij} \tag{3.9}$$

As shown in Figure 3.4, when the packing fraction of solid particles is higher, particle i may be in contact with several particles simultaneously; therefore, the force and moment imposed on the particle i are:

$$\vec{F}_i = \sum_j \left(\vec{F}_{nij} + \vec{F}_{tij}\right); \quad \vec{T}_i = \sum_j \left(a_i\vec{n} \times \vec{F}_{tij}\right) \tag{3.10}$$

3.1.2 Spring stiffness

The spring stiffness and damping coefficient introduced by the soft-sphere approach are related to the elastic modulus, Poisson's ratio, and various other parameters of particles, which cannot be measured directly and need to be calibrated. The normal spring stiffness k_n

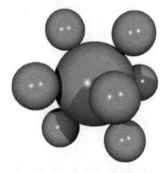

Figure 3.4: Schematics of multiple contacts.

can be determined using the Hertz contact theory:

$$k_n = \frac{4}{3}\left(\frac{1-v_i^2}{E_i}+\frac{1-v_j^2}{E_j}\right)^{-1}\left(\frac{a_i+a_j}{a_ia_j}\right)^{-1/2} \tag{3.11}$$

where E and v are the Young's approach and Poisson's ratio of particles, respectively; a is the particle radius; the subscripts i, j, respectively, represent the particles i and j in contact.

If the particles i and j have the same material properties and same size, then k_n is simplified into:

$$k_n = \frac{\sqrt{2a}E}{3(1-v^2)} \tag{3.12}$$

The tangential spring stiffness k_t is determined by the Mindlin–Deresiewicz contact theory:

$$k_t = 8\alpha^{1/2}\left(\frac{1-v_i^2}{G_i}+\frac{1-v_j^2}{G_j}\right)^{-1}\left(\frac{a_i+a_j}{a_ia_j}\right)^{-1/2} \tag{3.13}$$

where G_i and G_j are the shear modulii of particle i and particle j, respectively. If the particles i and j are comprised of the same material and the sizes are equal, k_t is simplified into:

$$k_t = \frac{2\sqrt{2a}G}{(2-v^2)}\alpha^{1/2} \tag{3.14}$$

Obviously, k_n and k_t are functions of the relative approaches, which need to be updated according to real-time contact process, but the computation is rather time-consuming. The soft-sphere approach usually assumes that the spring stiffness and damping coefficient remain constant during the entire contact process. A smaller value of spring stiffness leads to a larger calculation time step. In order to accelerate the calculation process, an unrealistically small value of spring stiffness is widely used in some large-scale engineering problems.

The analysis shows that in some liquid–solid systems, the simulation results obtained by using a small value of spring stiffness are consistent with the results obtained by using the precise value of spring stiffness. The interpretation may be that liquid–solid interactions decrease the impact of spring stiffness. Table 3.1 lists the specific values of parameters used in the soft-sphere approaches.

3.1.3 Damping coefficient

If the spring system, as illustrated in Figure 3.1, is at a critical damping state, the mechanical energy decays at the highest speed. Subsequently, the normal damping η_n and tangential damping η_t are expressed as:

$$\eta_n = 2\sqrt{mk_n} \tag{3.15}$$

$$\eta_t = 2\sqrt{mk_t} \tag{3.16}$$

Table 3.1: Parameter selection in soft-sphere approaches

Related work	k_n (kg/s^2)	η_n (kg/s)	e	μ
Tsuji et al.	800	0.018	0.9	0.3
Xu and Yu	50,000	0.15	0.9	0.3
Gera et al.	800	0.012	0.9	0.3
Kawaguchi et al.	800	0.012	0.9	0.1–0.3
Mikami et al.	800	0.0016	0.9	0.3
Xu et al.	40,000	0.12	0.9	–
Kuwagi et al.	800	8.26×10^{-6}	0.9	–
Rhodes et al.	800	0.0016	0.9	0.3
Xu et al.	50	1.65×10^{-5}	0.9	0.3
Limtrakul et al.	800	0.022	0.9	0.3

Note: k_n, η_n, e, and μ are the normal spring stiffness, damping coefficient, restitution coefficient, and friction coefficient, respectively.

Another method to determine the damping coefficient is to couple the spring coefficient with the restitution coefficient e:

$$\eta_n = -\frac{2 \ln e}{\sqrt{\pi^2 + (\ln e)^2}} \sqrt{mk_n} \qquad (3.17)$$

Note that e is determined experimentally. For $e = 0$, $\eta_n = 2\sqrt{mk_n}$. If the normal relative approach α is taken into account, η_n can be:

$$\eta_n = C\sqrt{mk_n}\alpha^{1/4} \qquad (3.18)$$

C is related to e; the relationship being shown in Figure 3.5. Determination of the tangential damping coefficient η_t is very complex, and usually takes the same value as η_n.

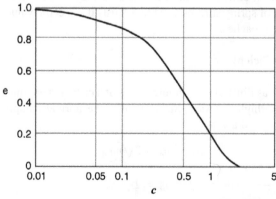

Figure 3.5: The relationship between parameters C and e in equation (3.18).

3.2 Hard-sphere approach

In a hard-sphere approach, the trajectories of particles are determined by momentum conservation during binary collisions. The interactions between particles are assumed to be instantaneous and additive, pair-wise. In the simulations, the collisions are processed one at a time, according to the order in which the events occur. For moderately dense systems, the hard-sphere models are considerably more efficient than the soft-sphere models. Note that the occurrence of multiple simultaneous collisions cannot be taken into account. The hard-sphere models have been applied to study a wide range of complex granular systems. Figure 3.6 indicates variations in contact potential energy in a dilute granular system. It can be seen that the contact duration, τ_{coll}, is much smaller in comparison with the free motion time, τ_c, between the two collisions. The hard-sphere approach is more suitable in such a system.

At high particle densities or at low coefficients of normal restitution (e), the collisions will lead to a dramatic decrease in kinetic energy. This is the so-called inelastic collapse; in this regime, the collision frequencies diverge as relative velocities disappear. Clearly, in such a case, the hard-sphere model is inappropriate.

3.2.1 One-dimensional collision

The simple case of a collision is the one-dimensional or head-on collision. Because of the conservation of momentum, the velocity of one of the particles after the collision can be exactly predicted. If particle 1 and particle 2 are subjected to a one-dimensional collision with each other, the relative velocity before collision is:

$$v_{12} \equiv v_1 - v_2 < 0 \tag{3.19}$$

For elastic collision, the energy is conserved, and the relative velocity after collision is:

$$v'_{12} = -v_{12} \tag{3.20}$$

Inelastic collisions may not conserve kinetic energy: $|v'_{12}| < |v_{12}|$.

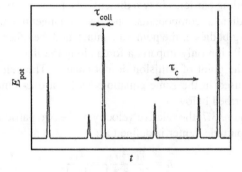

Figure 3.6: Variation of potential energy caused by particle collisions in a dilute granular system.

The ratio of the relative velocity of separation to the relative velocity of approach is defined as the coefficient of restitution, e:

$$e \equiv -v'_{12}/v_{12} \qquad (3.21)$$

Obviously, $0 \leq e \leq 1$. When $e = 1$, it represents a perfectly elastic collision, while $e = 0$ represents a perfectly inelastic collision.

The conservation of all momentum demands that the total momentum before the collision is the same as the total momentum after the collision, which is expressed by the equation based on momentum conservation:

$$m_1 v'_1 + m_2 v'_2 = m_1 v_1 + m_2 v_2, \qquad (3.22)$$

Together with equation (3.20), we have:

$$v'_1 - v'_2 = -e(v_1 + v_2), \qquad (3.23)$$

It can be derived that:

$$v'_1 = v_1 - \frac{m^*}{m_1}(1 + e)v_{12} \qquad (3.24)$$

$$v'_2 = v_2 - \frac{m^*}{m_2}(1 + e)v_{12} \qquad (3.25)$$

The effective mass m^* in equations (3.24) and (3.25) is defined as:

$$m^* \equiv \frac{m_1 m_2}{m_1 + m_2} \qquad (3.26)$$

For elastic collisions with bodies of equal mass, $m^*/m_1 = m^*/m_2 = 1/2$; their velocities after collision are:

$$v'_1 = v_2, \; v'_2 = v_1 \qquad (3.27)$$

3.2.2 Two- and three-dimensional collisions

In the case of two colliding particles in two dimensions, the velocity of each particle may be split into two perpendicular components: one that is tangential to the common normal surfaces of the colliding bodies at the point of contact, and the other that is along the line of collision. Since the collision only imparts a force along the line of collision, the velocities that are tangential to the point of collision do not change. The velocities along the line of collision can then be used in the same equations as for a one-dimensional collision. The derivations are as described below.

As illustrated in Figure 3.7, the relative velocity of the two particles is $\vec{v}_{12} = \vec{v}_1 - \vec{v}_2$, and the unit vector of the particle center is defined as:

$$\vec{n} = \frac{\vec{r}_1 - \vec{r}_2}{|\vec{r}_1 - \vec{r}_2|} = \frac{\vec{r}_{12}}{|\vec{r}_{12}|} \qquad (3.28)$$

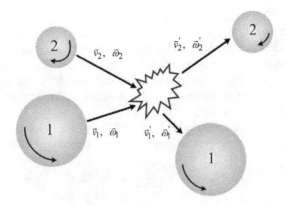

Figure 3.7: Schematics of a two-dimensional collision.

The normal component of the relative velocity is:

$$\vec{g} \equiv (\vec{v}_{12} \cdot \vec{n})\vec{n} = \vec{g}_{12}^n \tag{3.29}$$

The tangential component is:

$$\vec{g}_{12}^t \equiv \vec{v}_{12} - \vec{g}_{12}^n \tag{3.30}$$

Assuming the two particles do not rotate, the tangential component remains constant after the collision:

$$\left(\vec{g}_{12}^t\right)' = \vec{g}_{12}^t \tag{3.31}$$

Considering that the normal collision is equivalent to a one-dimensional collision, the normal velocity after collision is:

$$\vec{g}' = -e\vec{g}. \tag{3.32}$$

The velocities after collision become:

$$\vec{v}_1' = \vec{v}_1 - \frac{m^*}{m_1}(1 + e)(\vec{v}_{12} \cdot \vec{n})\vec{n} \tag{3.33}$$

$$\vec{v}_2' = \vec{v}_2 + \frac{m^*}{m_2}(1 + e)(\vec{v}_{12} \cdot \vec{n})\vec{n} \tag{3.34}$$

3.2.3 Normal restitution coefficient

The investigations into particle collisions can be traced back to Newton, whose original law of conservation of momentum was the basis for the impact theory. Its main assumption is that the objects are rigid, and the impact duration is zero. Since the law of conservation of momentum on its own is inadequate to determine the speed of each particle after the collision, the primary impact theory only considers two limiting cases: completely elastic collision and completely inelastic collision. Usually, the collision lies somewhere between these two extremes. The initial kinetic energy loss was realized through the introduction of the restitution coefficient, e, by Newton. The coefficient e is a comprehensive concept of energy loss, which may include different types of energy losses such as the viscoelasticity of materials, the plastic deformation of the contact surface, the vibration between two objects, and so on. Therefore, the restitution coefficient does not merely rely on any particular inherent attribute of materials; it also depends on the geometrical properties and impact speed of the contact surface.

Considering two particles with radii $R1$ and $R2$, when they contact each other with a relative approach α, the elastic contact force can be obtained by the Hertz contact theory:

$$F_{\text{Hertz}} = \frac{2Y\sqrt{R^*}}{3(1 - v^2)} \alpha^{3/2} \equiv B\alpha^{3/2} \tag{3.35}$$

where B is related to the material property of the particle. The dissipation force is inevitably related to α and the relative velocity $g = \dot{\alpha}$

$$F_{\text{diss}} = \frac{3}{2}A \cdot B\alpha^{1/2}\dot{\alpha} \tag{3.36}$$

where $A \equiv \frac{1}{3}\frac{(3\eta_2-\eta_1)^2}{(3\eta_2+2\eta_1)}\left[\frac{(1-v^2)(1-2v)}{Yv^2}\right]$ represents the dissipation characteristics of materials, which is the function of the respective viscous constants η_1 and η_2, while η_1 and η_2 both have relations with the dissipative stress tensor and deformation rate tensor.

The general relation of the elastic force and dissipative force is expressed as:

$$F_{\text{diss}} = A\dot{\alpha}\frac{\partial F_{\text{Hertz}}(\alpha)}{\partial\alpha} \tag{3.37}$$

This is applicable to smooth particles. Combining the two equations above, the total force imposed on the particles is:

$$\ddot{\alpha} + \frac{B}{m^*}\left(\alpha^{3/2} + \frac{3}{2}A\alpha^{3/2}\dot{\alpha}\right) = 0 \tag{3.38}$$

Setting the starting time of particle collision $t = 0$, the initial condition is:

$$\alpha(0) = 0, \quad \dot{\alpha}(0) = g \tag{3.39}$$

Here, $g = |\bar{g}|$ is the normal component of the instantaneous relative velocity when the particle collision occurs.

On the basis of the quasi-static assumptions in equation (3.37), it is obvious that the particle collision velocity is much smaller than the speed of propagation of sound in the particles. In other words, the viscous relaxation time is far less than the collision duration, τ_c. Thereby, the analytical solution for the velocity after collision can be obtained, and the closed form solution of e may be obtained as well:

$$e = 1 - C_1A\kappa^{2/5}g^{1/5} + C_2A^2\kappa^{4/5}g^{2/5} \mp \cdots \tag{3.40}$$

where:

$$\kappa \equiv \left(\frac{3}{2}\right)^{5/2}\left(\frac{B}{m^*}\right) = \left(\frac{3}{2}\right)^{3/2}\frac{Y\sqrt{R^*}}{m^*(1 - v^2)} \tag{3.41}$$

The parameters in equation (3.40) are:

$$C1 \approx 1.15344, \quad C2 \approx 0.79826, \quad C3 \approx -0.483582, \quad C4 \approx 0.285279 \qquad (3.42)$$

3.2.4 Tangential restitution coefficient

The tangential restitution coefficient e^t is related to the angular speed and tangential translational velocity. The particle velocity after collision can be determined by:

$$\left(\vec{g}^n\right)' = -e^n \, \vec{g}^n (0 \le e^n \le 1) \qquad (3.43)$$

$$\left(\vec{g}^t\right)' = -e^t \, \vec{g}^t (-1 \le e^t \le 1) \qquad (3.44)$$

where $\vec{g} = \vec{g}^n$ and $\vec{g}^t$ are the normal and tangential relative velocity, respectively, of the contact point before collision. If particle size is equal, we have:

$$\vec{g}^n = (\vec{v}_{12} \cdot \vec{n})\vec{n} = \vec{g} \qquad (3.45)$$

$$\vec{g}^t = \vec{v}_{12} - \vec{g} + R[\vec{n} \times (\vec{\omega}_1 + \vec{\omega}_2)] \qquad (3.46)$$

where $\vec{\omega}_1$ and $\vec{\omega}_2$ are the angular speeds of particles 1 and 2, respectively. When $e^t = 1$, the tangential motion is constant. When $e^t = -1$, the tangential motion reverses.

If considering particle rotations, the angular speed $\vec{\omega}_1$ obeys:

$$\frac{dI\vec{\omega}_1}{dt} = \vec{r} \times \vec{F}_{12} \qquad (3.47)$$

where I is the moment of inertia; F_{12} is the contact force; $\vec{r} = -R\vec{n}$ is the position vector of particle 1. Assuming that the relative approach of the particle during the collision process is

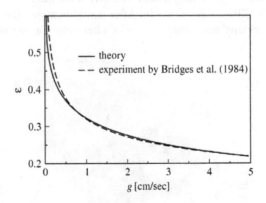

Figure 3.8: The effect of ice particle collision velocity on the normal restitution coefficient. When the collision velocity is small, the calculated data are different from the experimental results. This is primarily because the adhesion forces between ice particles are relatively large.

small compared with the particle size, equation (3.47) becomes:

$$I(\vec{\omega}_1' - \vec{\omega}_1) = -R[\vec{n} \times m(\vec{v}_1' - \vec{v}_1)] \tag{3.48}$$

$$I(\vec{\omega}_2' - \vec{\omega}_2) = -R[\vec{n} \times m(\vec{v}_2' - \vec{v}_2)] \tag{3.49}$$

$$\vec{v}_1' + \vec{v}_2' = \vec{v}_1 + \vec{v}_2 \tag{3.50}$$

Equation (3.50) is the conservation of translational momentum, which can then derive the particle velocity after the collision.

$$\vec{v}_1' = \vec{v}_1 - \frac{1+e^n}{2}\vec{g}^n - \frac{1+e^t}{2(1+q^{-1})}\vec{g}^t \tag{3.51}$$

$$\vec{v}_2' = \vec{v}_2 + \frac{1+e^n}{2}\vec{g}^n + \frac{1+e^t}{2(1+q^{-1})}\vec{g}^t \tag{3.52}$$

$$\vec{\omega}_1' = \vec{\omega}_1 + \frac{1-e^t}{1+q}\frac{1}{2R}\left[\vec{n} \times \vec{g}^t\right] \tag{3.53}$$

$$\vec{\omega}_2' = \vec{\omega}_2 + \frac{1-e^t}{1+q}\frac{1}{2R}\left[\vec{n} \times \vec{g}^t\right] \tag{3.54}$$

Here $q \equiv I/(mR^2)$; its physical meaning is the effective moment of inertia. When $e^t = 1$, equations (3.51) and (3.52) can be simplified into equations 3.33 and 3.34, respectively.

The tangential motion of a particle is difficult to determine. It can be seen from Figure 3.9 that there are two different types of tangential motion. If g^n is smaller and g^t is larger, and $e^t > 0$, particles slip relative to each other, since a small normal velocity induces a small normal force and small tangential force. If g^n is larger and g^t is smaller, the normal force is large enough that the imposed tangential force may force the two particles into reversing their rotation after collision, so $e^t < 0$. After simplification, the analytical solutions of the translational velocity and angular velocity after collision are obtained, as shown in Table 3.2.

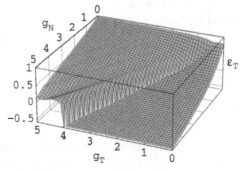

Figure 3.9: The tangential restitution coefficient as a function of normal velocity and tangential velocity.

Table 3.2: Solutions of velocity after collision in the hard-sphere approach

Conditions	$\dfrac{\bar{n}\cdot\bar{g}}{|\bar{g}_{ct}|} < \dfrac{2}{7(1+e)}$	$\dfrac{\bar{n}\cdot\bar{g}}{|\bar{g}_{ct}|} > \dfrac{2}{7(1+e)}$
Translational speed	$\bar{v}'_1 = \bar{v}_1 - (\bar{n} - \mu\bar{t})(\bar{n}\cdot\bar{g})(1+e)\dfrac{m_2}{m_1+m_2}$	$\bar{v}'_1 = \bar{v}_1 - \left\{(1+e)(\bar{n}\cdot\bar{g})\bar{n} + \dfrac{2}{7}|\bar{g}_{ct}|\bar{t}\right\}\dfrac{m_2}{m_1+m_2}$
	$\bar{v}'_2 = \bar{v}_2 + (\bar{n} - \mu\bar{t})(\bar{n}\cdot\bar{g})(1+e)\dfrac{m_1}{m_1+m_2}$	$\bar{v}'_2 = \bar{v}_2 - \left\{(1+e)(\bar{n}\cdot\bar{g})\bar{n} + \dfrac{2}{7}|\bar{g}_{ct}|\bar{t}\right\}\dfrac{m_1}{m_1+m_2}$
Angular speed	$\bar{\omega}'_1 = \bar{\omega}_1 + \left(\dfrac{5}{2a_1}\right)(1+e)(\bar{n}\cdot\bar{g})(\bar{n}\times\bar{t})\dfrac{m_2}{m_1+m_2}$	$\bar{\omega}'_1 = \bar{\omega}_1 - \left(\dfrac{5}{7a_1}\right)|\bar{g}_{ct}|(\bar{n}\times\bar{t})\dfrac{m_2}{m_1+m_2}$
	$\bar{\omega}'_2 = \bar{\omega}_1 + \left(\dfrac{5}{2a_2}\right)(1+e)(\bar{n}\cdot\bar{g})(\bar{n}\times\bar{t})\dfrac{m_2}{m_1+m_2}$	$\bar{\omega}'_2 = \bar{\omega}_2 - \left(\dfrac{5}{7a_2}\right)|\bar{g}_{ct}|(\bar{n}\times\bar{t})\dfrac{m_2}{m_1+m_2}$

Note: e is the normal restitution coefficient, μ is the surface friction coefficient, and $\bar{g} = \bar{v}_1 - \bar{v}_2$ is the particle relative velocity before collision. At the point of collision, $\bar{g}_{ct} = \bar{g} + a_1\bar{\omega}_1 \times \bar{n} + a_2\bar{\omega}_2 \times \bar{n}$ is the relative velocity of the particle before the collision. The variable $\bar{n}$ is the normal unit vector and $\bar{t} = \bar{g}_{ct}/|\bar{g}_{ct}|$ is the tangential unit vector.

3.3 Comparisons

Both hard- and soft-sphere models have been used in the simulation of granular materials, and each model has its own characteristics. This makes each of the two models very efficient in some cases and inefficient in other cases.

The soft-sphere approach simplifies the contact force between the particles by using the spring stiffness and the damping coefficient, which normally assumes that the parameters are unchanged in the contact process. This ignores the details such as loading history, etc., and the contact force can be calculated directly using the relative approach between the particles. Therefore, the calculation is obviously less robust. Furthermore, the elastic and damping coefficients are related to physical parameters such as the measurable particle elastic modulus and Poisson's ratio, etc.; a reasonable calibration is the key to whether the soft-sphere approach is accurate or not. In the soft-sphere approach, the normal spring stiffness k_n needs to be determined at first; the tangential spring stiffness k_t is correlated with k_n. In many simulations, especially those involving large amounts of particles, smaller values of k_n and k_t are used, but only in order to accelerate the computations. Therefore, some non-physical mechanisms are certainly introduced.

In Table 3.3, the two types of models are compared. Hard-sphere models use an event-driven scheme because the interaction times are assumed to be small compared to the free flight time of particles, where the progression in physical time depends on the number of collisions that occur. On the other hand, when considering soft-sphere models, a time step that is significantly shorter than the contact time should be used. This directly implies that the computational efficiency of the soft-sphere model decreases when the ratio of the free-flight time to the contact time increases. This is the case when the system becomes less dense. In the soft-sphere models, a slight deformation of particles is allowed so that the multiple contacts between several pairs of particles are possible. This should never happen in the event-driven models. As mentioned above, a lower coefficient of normal restitution may lead to the inelastic collapse observed in hard-sphere simulations. Incorporation of cohesive forces, especially pair-wise forces, is quite straightforward in soft-sphere models. This is because the collision process in the soft-sphere model is described via the Newtonian

Table 3.3: Comparison between three types of approaches

Physical quantity	Soft-sphere approach	Hard-sphere approach
Time step	Dependent on contact time	–
Velocity	Based on contact force	Given directly
Preciseness	Simplified	Over-simplified
Computing efficiency	Low	Very high
Input parameter	Stiffness and damping coefficient that need to be calibrated	Restitution coefficient, etc., that need to be calibrated
Application scope	Engineering applications, dense systems	Specific applications in dilute systems

equations of motion of individual particles, that is, in terms of forces. In the hard-sphere system, the update is not through forces, but through an exchange of momentum at contact. This indicates that for short-range forces, such as the cohesive force, a type of hybrid method for the interaction at close encounters has to be devised. This is not straightforward. On the other hand, for systems with different size particles, it is the soft-sphere model which poses some obstacles. In a soft-sphere system using a linear-spring/dashpot scheme, the value of the spring stiffness is dependent on the particle size. This means that in principle, a different value of spring stiffness should be used for calculating the contact forces between particles with different sizes. Otherwise, the computing efficiency will drop substantially.

References

[1] N. Brilliantov and T. Pöschel, *Kinetic Theory of Granular Gases*, Oxford University Press, UK (2004).

[2] C. T. Crowe, *Multiphase Flow Handbook*, CRC Press, Taylor & Francis Group, USA (2005).

[3] P. A. Cundall and O. D. L. Strack, 'A discrete numerical model for granular assembles', *Geotechnique*, 29(1), 47–65 (1979).

[4] D. Gera, M. Gautam, Y. Tsuji, T. Kawaguchi and T. Tanaka, 'Computer simulation of bubbles in large-particle fluidized beds', *Powder Technol.*, 98(1), 38–47 (1998).

[5] K. Kuwagi, T. Mikami and M. Horio, 'Numerical simulation of metallic solid bridging particles in a fluidized bed at high temperature', *Powder Technol.*, 109(1–3), 27–40 (2000).

[6] T. Kawaguchi, T. Tanaka and Y. Tsuji, 'Numerical simulation of two-dimensional fluidized bed using the discrete element method (Comparison Between the Two- and Three-dimensional Model)', *Powder Technol.*, 96(2), 129–138 (1998).

[7] S. Limtrakul, A. Chalermwattanatai, K. Unggerawirote, Y. Tsuji, T. Kawaguchi and W. Tanthapanichakoon, 'Discrete particle simulation of solids motion in a gas-solid fluidized bed', *Chem. Eng. Sci.*, 58(3–6), 915–921 (2003).

[8] T. Mikami, H. Kamiya and M. Horio, 'Numerical simulation of cohesive powder behavior in a fluidized bed', *Chem. Eng. Sci.*, 53(10), 1927–1940 (1998).

[9] T. Pöschel and T. Schwager, *Computational Granular Dynamics: Models and Algorithms*, Springer-Verlag, Heidelerg, Germany (2005).

[10] M. J. Rhodes, X. S. Wang, M. Nguyen, P. Stewart and K. Liffman, 'Onset of cohesive behavior in gas fluidized beds: a numerical study using DEM simulation,' *Chem. Eng. Sci.*, 56, 4433–4438 (2001).

[11] Y. Tsuji, T. Kawaguchi and T. Tanaka, 'Discrete particle simulation of two-dimensional fluidized bed', *Powder Technol.*, 77(1), 79-87 (1993).

[12] M. W.Weber, 'Simulation of cohesive particle flows in granular and gas–solid system', Doctoral dissertation, University of Colorado at Boulder, USA (2004).

[13] B. H. Xu and A. B. Yu, 'Numerical simulation of the gas-solid flow in a fluidized bed by combining discrete particle method with computational fluid dynamics', *Chem. Eng. Sci.*, 52(16), 2785–2809 (1997).

[14] B. H. Xu, Y. C. Zhou, A. B. Yu and P. Zulli, 'Force structures in gas fluidized beds of fine powders', *World Congress on Particle Technology 4*, Sydney, Australia (2002).

Chapter 4

Liquid bridge forces

Liquid bridges may be established between the surfaces of wet particles when they are in proximity to each other, which then adheres the particles. The mechanical properties of granular systems may therefore change significantly. At the present stage, the mechanisms behind the liquid bridge are poorly understood. The studies on the mechanics of granular matter predominantly focus on dry assemblies. The humidity in lab experiments is reduced as much as possible to avoid the adhesion induced by liquid bridges, which can therefore simplify the analysis of the behavior of granular matter systems. In nature, wet particles are prevalent, such as unsaturated soil. The deformation and strength of this soil are closely related with the level of saturation, and are more complex to predict than those of fully saturated soil. A typical example is the Kansai Airport, constructed in 1987 in Osaka, Japan, which is situated on an artificial island in Osaka Bay. This site was built with 0.18 billion cubic meters of earth; due to inadequate consideration of the soil water content, the airport has sunk 11.5 meters (as of 2000). Therefore, wet granular mechanics is extremely important in engineering applications.

4.1 Liquid distribution

Dry granular matter shows sophisticated phenomena. If some liquid is added to increase the adhesion, it is bound to increase the complexity of the behavior of the granular matter. For example, the angle of repose of a pile of sand is maintained at about 30°, which has nothing to do with the magnitude of the sand pile. When an appropriate amount of water is added to the pile, there exists adhesion between the particles of sand, and the angle of repose may even exceed 90°, as shown in Figure 4.1.

The shape of the liquid bridge is primarily dependent on the content of the interstitial fluid, which in turn is usually expressed by the saturation s, that is, the percentage of volume occupied by the liquid, as shown in Table 4.1.

Liquid bridge forces can be classified into two types: static and dynamic. The static liquid bridge force is formed by the joint action of the negative pressure difference and the tension on the liquid surface. When there is a relative motion of two particles, the viscous interstitial fluid generates a normal force and a tangential shear resistance on the particle surface, and these two forces constitute the dynamic liquid bridge force. The static liquid bridge force is

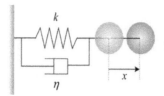

Figure 4.1: Sand piles stacked with dry and wet particles of sand.

related to the geometry of the liquid bridge, while the dynamic liquid bridge force is related to particle motion and fluid viscosity.

The pendular liquid bridge, with a low saturation, is easy to test and analyze. A dimensionless parameter, C_a, is often used to measure the ratio of the dynamic liquid bridge force to the static liquid bridge force:

$$C_a = \frac{\mu v}{\gamma} \tag{4.1}$$

At 20°C, the surface tension coefficient of water $\gamma = 72.75 \times 10^{-3}$ N/m, and viscosity $\mu = 1.0 \times 10^{-4}$ Pa/s. If the relative velocity V between particles is about 10 m/s, then $C_a < 0.01$. Under this circumstance the dynamic effect can be ignored and only the static liquid bridge force between particles is considered. When the relative velocity between particles or the interstitial fluid adhesion is greater, the adhesion generated by the dynamic liquid bridge force plays a prominent role.

4.2 Static liquid bridge force

The liquid present on wet particles generally forms a thin film and evenly covers the surfaces of these particles. When the two particles are close to each other, some of the film begins to merge together, due to a large and negative surface curvature. All the liquid in the other film is drawn to this overlap region and gradually forms a stable liquid bridge, as shown in Figure 4.2.

The capillary length l_c can be used to measure the relative magnitudes of the liquid bridge force and particle gravity:

$$l_c \equiv \sqrt{\frac{\gamma}{\rho g}} \tag{4.2}$$

Where, g is the acceleration due to gravity, and ρ the fluid density. The capillary length of pure water at room temperature is about 2.7 mm, which indicates that for particles with a smaller particle size, the liquid bridge force plays a major role.

The contact angle can also be considered in terms of the thermodynamics of the materials involved. By denoting the solid–vapor interfacial energy (see surface energy) as γ_{sv}, the solid–liquid interfacial energy as γ_{sl}, and the liquid–vapor energy (i.e., the surface tension) as simply γ, the contact angle θ must be satisfied in equilibrium (known as

Table 4.1: The interstitial liquid between particles and corresponding liquid bridge shapes

Saturation, s	Liquid bridge morphology	Schematic diagram	Physical mechanism
0	–		No liquid bridge force
<30%	Pendular		Liquid bridges are very thin. Particles adhere to each other through the liquid bridges on the contact points
30% < s < 70%	Funicular		A liquid bridge exists in the surroundings of the contact point. Parts of the gaps between particles are fully filled with fluid, and cohesion increases
>70%	Capillary		Liquid almost fully fills each gap. The gas pressure is greater than the liquid pressure, which depresses the liquid surface and generates suction between particles
100%	Slurry		Liquid pressure is equal to or greater than the air pressure. There is no cohesion force between particles

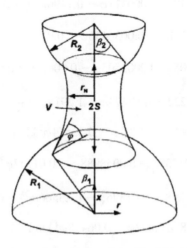

Figure 4.2: A schematic representation of a pendular liquid bridge of volume V and surface tension γ between two spheres of radii R_1 and R_2, respectively, separated by a distance $2S$, a liquid–solid contact angle φ, and half-filling angles β_1 and β_2, respectively.

the Young equation):

$$\cos \theta = \frac{\gamma_{sg} - \gamma_{sl}}{\gamma} \tag{4.3}$$

Where γ_{sg} and γ_{sl} are the free energies per unit area of the solid–liquid and solid–gas interface, respectively, and γ the liquid surface tension. When $\gamma_{sg} > \gamma_{sl}$, then the contact angle $\theta < 90°$, and the solid surface is partially wet. If $\gamma_s = 0$, the liquid has completely wetted the surfaces of the particles, and the contact angle is $0°$.

The static liquid bridge force is the sum of the surface tension and hydrostatic pressure.

$$F_s = 2\pi\gamma r_N - \pi r_N^2 \Delta p \tag{4.4}$$

The former is the component of surface tension in the direction of the particle center; while the latter is the hydrostatic pressure force inside and outside the gas–liquid interface. r_N is the radius of curvature of the liquid bridge neck; Δp is the hydrostatic pressure difference between the gas–liquid interfaces of the liquid bridge. This parameter Δp is negative, generating tension between the two particles. The surface tension and hydrostatic pressure forces are both area-distributed forces, which have the same active area and direction and generate attraction between particles.

A considerably more complex expression is required for greater volumes and contact angles, and this is presented as:

$$\ln F^* = f_1 - f_2 \exp\left(f_3 \ln S^+ + f_4 \ln^2 S^+\right) \tag{4.5}$$

In which:

$$f_1 = \left(-0.44507 + 0.05083\varphi - 1.1466\varphi^2\right) +$$
$$\left(-0.1119 - 0.000411\varphi - 0.1490\varphi^2\right) \ln V^* +$$
$$\left(-0.012101 - 0.0036456\varphi - 0.01255\varphi^2\right)\left(\ln V^*\right)^2 +$$
$$\left(-0.0005 - 0.0003505\varphi - 0.00029076\varphi^2\right)\left(\ln V^*\right)^3 \tag{4.6}$$

$$f_2 = \left(1.9222 - 0.57473\varphi - 1.2918\varphi^2\right) +$$
$$\left(-0.0668 - 0.1201\varphi - 0.22574\varphi^2\right) \ln V^* +$$
$$\left(-0.0013375 - 0.0068988\varphi - 0.01137\varphi^2\right)\left(\ln V^*\right)^2 \tag{4.7}$$

$$f_3 = \left(1.268 - 0.01396\varphi - 0.23566\varphi^2\right) +$$
$$\left(0.198 + 0.092\varphi - 0.06418\varphi^2\right) \ln V^* +$$
$$\left(0.02232 + 0.02238\varphi - 0.009853\varphi^2\right)\left(\ln V^*\right)^2 +$$
$$\left(0.0008585 + 0.001318\varphi - 0.00053\varphi^2\right)\left(\ln V^*\right)^3 \tag{4.8}$$

$$f_4 = \left(-0.010703 + 0.073776\varphi - 0.34742\varphi^2\right) +$$

$$\left(0.03345 + 0.04543\varphi - 0.09056\varphi^2\right) \ln V^* + \qquad (4.9)$$

$$\left(0.0018574 + 0.004456\varphi - 0.006257\varphi^2\right) \left(\ln V^*\right)^2$$

The expression is valid for $\varphi < 0.873 (\sim 50°)$ and $V^* < 0.1$ and gives an error in the force estimate of less than 3%. Using radians for the contact angle φ, we have given the average radius of curvature R on the liquid bridge neck, and the other dimensionless parameters by:

$$\frac{1}{R} = \frac{1}{2} \left(\frac{1}{R_1} + \frac{1}{R_2} \right) \qquad (4.10)$$

$$V^* = \frac{V}{R^3} \qquad (4.11)$$

$$F^* = \frac{F}{2\pi R\gamma} \qquad (4.12)$$

$$S^* = \frac{S}{R} \qquad (4.13)$$

$$S^+ = \frac{S^*}{\sqrt{V^*}} \qquad (4.14)$$

It can be seen from Figure 4.3 that saturation has a large influence on the hydrostatic pressure. In particular, the hydrostatic pressure force changes significantly in the pendular state and capillary state. The nonconformity between the wetting and drying processes in Figure 4.3 is caused by the different liquid film thicknesses in the formation of the meniscus in the pore.

It can be seen in Figure 4.4 that the cohesive stress can be accurately measured in the pendular and capillary states. It increases along with the increase in saturation in the pendular state, while it decreases rapidly with saturation in the capillary state. In the funicular state, the cohesive stress undergoes large fluctuations because the internal configurations of particles and the interstitial liquid vary widely under this level of saturation.

4.2.1 Separation distance

Although the force exerted by the liquid bridges between particles is weak, an increase of both the relative velocity and the acceleration of the contact frequency between particles can create a situation in which the frequent formation and destruction of the liquid bridges between particles can certainly have a significant influence on the mechanical behavior of wet granular matter. Therefore, it is essential to describe the stability of liquid bridges.

When the gap between particles is smaller than a critical value, the liquid bridge is stable. Therefore, this length of the liquid bridge can be referred to as the critical separation

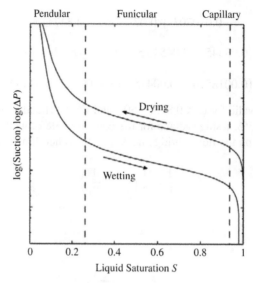

Figure 4.3: Variation of hydrostatic pressure force with saturation in equation (4.4).

distance S_c. The Laplace–Young equation is directly related to S_c. Unfortunately, there is no satisfactory theoretical explanation for the judgment of S_c. For a nonadhesive liquid, when the contact angle $\theta < 40°$, the relationship between S_c and the liquid bridge volume V is:

$$S_c = (1 + 0.5\theta)V^{1/3} \tag{4.15}$$

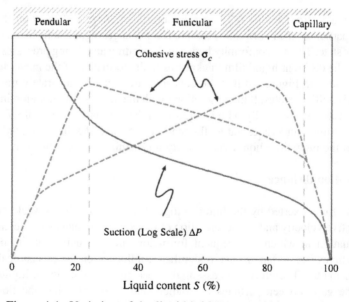

Figure 4.4: Variation of the liquid bridge stress with saturation.

Where, the contact angle θ is calculated using radians, for $0 < \theta < 0.70 (\sim 40°)$. If $\theta = 0$, it becomes:

$$S_c = V^{1/3} \tag{4.16}$$

The stability of the liquid bridge is not only related to its volume, but also to the motion of the fluid and its surface tension. Taking into account the liquid viscosity μ, the surface tension coefficient γ, and the relative velocity v_{re} between particles, we can express S_c as:

$$S_c = \left(1 + \sqrt{C_a}\right)(1 + 0.5\theta)V^{1/3} \tag{4.17}$$

Under conventional experimental conditions, the surface tension of the liquid is dominant and C_a is small; therefore, equation (4.17) can be reduced to equation (4.15).

4.2.2 Multiple liquid bridge force

In experimental or theoretical analysis, the mass or volume ratios of the interstitial fluid to those of particles is often known; the static liquid bridge force of internal individual particles has to be calculated. As shown in Figure 4.5, the mass ratio of the interstitial fluid to a dry particle in wet granular material is given as:

$$w_0 = \frac{W_1}{W_p} = \frac{n_c \rho_l V_1}{2} \frac{1}{\rho_p} \left(\frac{4}{3}\pi R_p^3\right)^{-1} = \frac{3n_c \rho_l}{8\pi \rho_p R_p^3} \cdot V_1 \tag{4.18}$$

Where, V_1 is the volume of a single liquid bridge, W_1 and W_p the masses of the interstitial liquid and the dry particles, respectively; ρ_l and ρ_p the density of the interstitial liquid and the particle, respectively; n_c the average number of particles in contact with a single particle, that is, the coordination number.

A single liquid bridge volume V_1 is expressed as:

$$V_1 = \frac{8\pi \rho_p R_p^3 w_0}{3n_c \rho_l} \tag{4.19}$$

Substituting equation (4.19) to (4.5), the single liquid bridge force can be calculated. For a different distribution of particles, n_c would vary. Regarding the simple cubic structure composed of equally sized particles, $n_c = 8$ and the joint force of the static liquid bridge forces imposed on the particles can be obtained, as schematically shown in Figure 4.5.

Figure 4.5: The liquid bridge force imposed on the particle i among wet particles.

4.3 Dynamic liquid bridge force

Dynamic liquid bridge forces between particles are primarily caused by the viscosity of the interstitial fluid. For the pendular liquid bridge, the impact of the viscosity of the liquid is reflected in two aspects. First, the liquid viscosity affects the evolution of a concave gas–liquid surface shape, the area proximal to this surface is known as the nonlubricated area; second, the liquid viscosity controls the strength of the liquid bridge, whereby this area is known as the lubricated area. This also affects the shape of the gas–liquid interface. When the gap between two particles is very small, the contribution of the gas–liquid interface to the dynamic liquid bridge force is negligible, and thus, the lubrication theory is used to analyze the dynamic liquid bridge force.

As illustrated in Figure 4.6, two spheres of radii R_1 and R_2 move toward each other along their common axis with a relative velocity V. s_0 is the minimum separation distance at the central axis of the two surfaces. B is the upper limit in the radial direction. In the case of a finite liquid bridge, B is the radius of the liquid bridge. For fully immersed spheres, B corresponds to the point where the pressure approaches the bulk value. For a small gap, the separation distance between the two proximate surfaces S_1 and S_2 is approximated by

$$s(r) = s_0 + \frac{r^2}{2R^*} \tag{4.20}$$

Where R^* is the effective radius.

$$R^* = \frac{R_1 R_2}{R_1 + R_2} \tag{4.21}$$

4.3.1 Normal force of Newtonian fluid

Elasto-hydrodynamic lubrication is the study of elastically deforming lubricated surfaces. When the Young's modulus of particles is large, the particle deformation can be ignored.

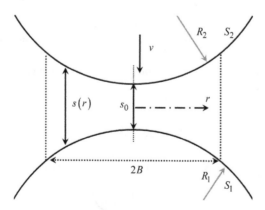

Figure 4.6: Schematic diagram of the squeeze flow between two rigid spheres, showing the coordinate system used.

When the interstitial fluid is Newtonian, the normal component of the viscous force between two deformable solids may only be obtained by elaborate numerical procedures. However, an approximate closed-form solution has been developed. For relatively hard spheres, the analytical solution degenerates to the rigid sphere solution, which formulates the viscous force F_n in the normal direction of approach as:

$$F_n = 6\pi Kv(R^*)^2/s_0 \qquad (4.22)$$

Where K is the viscosity, v the relative normal velocity between the two spheres, and R^* the effective radius expressed in terms of the radii of the two spheres. When the particle gap s_0 approaches 0, the force would have a singular value. For an elastic ball, the elastic deformation on the contact surface will occur in tandem with an increase in radial pressure. There will exist a small separation distance, such as 0.1% of the particle size. A large number of calculations show that for relatively hard particles, the calculation results obtained with equation (4.22) show good stability.

4.3.2 Normal force of power-law fluid

Since the early work on power-law fluids by Scott, many studies have been conducted on the squeeze flow of non-Newtonian fluids between rigid parallel plates with or without wall-slip. There have been relatively few studies involving rigid spherical platens or particles. Adams and Edmondson formulated a solution to the lubrication problem for a power-law fluid between two equal spheres, but they did not obtain a closed-form solution for the viscous force. Rodin considered nearly touching unequally sized spheres embedded in a power-law fluid and argued that the viscous resistance depends only on the inner, rather than the outer, gap region between the spheres when the flow index, n, is greater than 1/3; these regions will be discussed in more detail later. He derived an asymptotic solution that requires numerical evaluation. It predicts a singularity in the viscous force when $n \to 1/3$, which is physically unreasonable. Furthermore, he was unable to obtain a solution for $n < 1/3$.

For the squeeze flow with no-slip condition, it can be indicated that when the flow index n is greater than 1/3 and the gap is sufficiently small, the pressure will be concentrated in the inner region of the gap. Thus, the upper limit B can be chosen in the outer region such that $B = \min(R_1, R_2)$.

Under the condition of a sufficiently small gap between the two spheres, the inertial force has a negligible effect. The governing equations for the squeeze flow can be simplified using the Reynolds lubrication analysis. For a power-law fluid, the shear stress is related to the shear rate by the following constitutive equation:

$$\sigma = K\gamma^n \qquad (4.23)$$

Where, n is the flow index, K the constant coefficient, and γ the shear rate. As $n = 1$, the fluid is Newtonian.

The following closed-form approximation for the viscous force is:

$$F_n = 2\pi K (R^*)^2 \left(\frac{2n+1}{n}\right) \left(\frac{R^*}{s_0}\right)^{(3n-1)/2} \left(\frac{V}{R^*}\right)^n \times$$

$$\begin{cases} 3e^{-2.4n} + \dfrac{3}{2}e^{-1.06n} + 2^{5/3}\left[\ln\dfrac{b}{4} + \dfrac{5}{3}\left(\dfrac{1}{b^2} - \dfrac{1}{16}\right)\right], & n = 1/3 \\[3mm] 3e^{-2.4n} + \dfrac{3}{2}e^{-1.06n} + 2^{(3+n)/2}\left[\dfrac{b^{1-3n} - 4^{1-3n}}{1 - 3n} + \dfrac{4n+2}{3n+1}\left(\dfrac{1}{b^{1-3n}} - \dfrac{1}{4^{1-3n}}\right)\right], & n \ne 1/3 \end{cases} \qquad (4.24)$$

In which $b = B/\sqrt{R^* s_0}$.

For $n = 1$, equation (4.24) is reduced to $F_n = 6\pi K v (R^*)^2/s_0$, which is identical to the analytical solution for the squeeze flow of a Newtonian fluid; see equation (4.22).

In Figure 4.7, the dimensionless viscous force $\overline{F} = F/\pi K(R^*)^{2-n}v^n$ is plotted as a function of the flow index and compared with the asymptotic solution for a range of dimensionless separation distances, s_0/R^*. Note that $B = 20\sqrt{R^* s_0}$. It may be seen that there is a close agreement between the two equations when $n > 0.6$. However, for $n < 0.6$ the asymptotic solution overestimates the viscous force and, as mentioned previously, exhibits singularity when n approaches $1/3$, which was not predicted by equation (4.24).

Figure 4.8 is for a wider range of B, i.e., $B = 0.2R^*$ and solid lines to $B = R^*$.

It can be seen that equation (4.24) is accurate for $n > 0.5$ and is in close agreement with a previously published asymptotic solution in the range $n > 0.6$. The two closed-form solutions show continuous and monotonic behavior for all values of n. Further, the solution satisfying the lubrication limit is valid in the range $n < 1/3$, provided that it is restricted to liquid bridges.

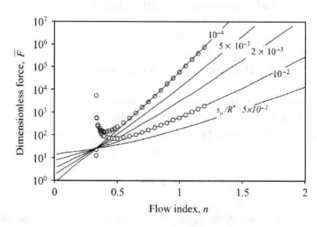

Figure 4.7: The dimensionless viscous force predicted by the approximate analytical expression for ($B = 20R^*$) (solid lines) compared with Rodin's asymptotic solution (circles) for a range of dimensionless separation distances.

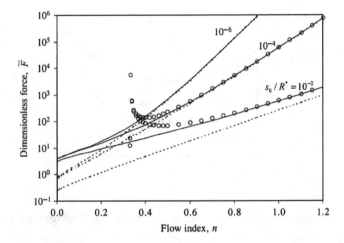

Figure 4.8: The dimensionless viscous force predicted by the approximate analytical expression, (solid lines) compared with Rodin's asymptotic solution (circles) for a range of dimensionless separation distances. Dotted lines correspond to $B = 0.2R*$ and solid lines to $B = R*$.

4.3.3 Tangential resistance of Newtonian fluid

Tangential resistance is characterized by the axial asymmetry and has been difficult to predict so far. Goldman et al. [13] studied the simple case of rigid balls parallel to rigid walls in a semi-infinite Newtonian fluid. When the separation distance between the ball and wall is sufficiently small ($s_0 \rightarrow 0$), the asymptotic solution of the tangential adhesion force is:

$$F = 6\pi KR*u\left(\frac{8}{15}\ln\frac{R^*}{s_0} + 0.9588\right) \qquad (4.25)$$

Where, u is the tangential relative velocity.

4.3.4 Tangential force of power-law fluid

The analysis of the tangential forces between particles with interstitial power-law fluid is very complex. Xu et al. derived the numerical solutions of the approximate pressure equation, resistance, and resistance moment, under additional assumptions of the velocity field by the small-parameter method. They proved that when the power exponent is 1, it can be re-treated as Goldman's asymptotic solution of Newtonian fluid, thereby avoiding its singularity.

To analyze the tangential interaction, consider two rigid spheres of different radii, R_1 and R_2, with a gap $s_0 \ll \min(R_1, R_2)$. The upper sphere translates perpendicularly along the common axis with a tangential velocity u, as shown in Figure 4.9. In this case the problem is nonsymmetrical.

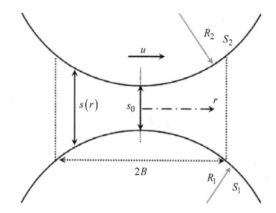

Figure 4.9: The tangential resistance of the interstitial fluid between particles.

Assuming $p^* = \dfrac{p_0(r)s_0^{n+1}}{12nKBU^n}$, $\bar{r} = \dfrac{r}{B}$, $c = \dfrac{B^2}{2R^*s_0}$, $\bar{s}(\bar{r}) = \dfrac{s(r)}{s_0} = 1 + c\bar{r}^2$, $\lambda = \dfrac{2R^*}{R_1}$, then
the pressure distribution along the particle surface is:

$$\frac{d^2 p^*}{d\bar{r}^2} + \left[2c(n+2)\frac{\bar{r}}{\bar{s}} + \frac{1}{\bar{r}}\right]\frac{dp^*}{d\bar{r}} + \left[\frac{2c(n+2)(1-n)}{(n+3)\bar{s}} - \frac{1}{\bar{r}^2}\right]p^* = -\frac{4c(\lambda-1)}{(n+3)}\frac{\bar{r}}{\bar{s}^{n+2}} \quad (4.26)$$

Assuming $F_0 = 2c\pi KR^*\bar{u}^n s_0^{1-n}$, $M_0 = F_0 R_1$, the horizontal resistance F_x on the spheres is:

$$F_x = -F_0 f_x = -F_0 \int_0^1 \left(12nc\lambda p^* \bar{r}^2 + \frac{2\bar{r}}{\bar{s}^n} + 6\bar{s}\bar{r}\frac{\partial p^*}{\partial \bar{r}} + 6\bar{s}p^*\right)d\bar{r} \quad (4.27)$$

The moment M_y exerted on the upper sphere by the pressure can be calculated by:

$$M_y = M_0 m_y = M_0 \int_0^1 \left(\frac{2\bar{r}}{\bar{s}^n} + 6\bar{s}\bar{r}\frac{\partial p^*}{\partial \bar{r}} + 6\bar{s}p^*\right)d\bar{r} \quad (4.28)$$

The numerical solutions of the dimensionless resistance and resistance moment are given by:

$$f_x = \frac{3n+3\lambda-1}{c(n-1)(3n+2)}\left[1 - \frac{1}{(1+c)^{n+1}}\right] + \frac{3(\lambda-1)(n\lambda-n-1)}{cn(n-1)(3n+2)}\left[1 - \frac{1+cn}{(1+c)^n}\right] \quad (4.29)$$

$$m_y = \frac{3n+3\lambda-1}{c(n-1)(3n+2)}\left[1 - \frac{1}{(1+c)^{n+1}}\right] - \frac{3(\lambda-1)(n+1)}{cn(n-1)(3n+2)}\left[1 - \frac{1+cn}{(1+c)^n}\right] \quad (4.30)$$

When the sphere moves in a parallel direction to the flat wall, $\lambda = 2$ and $B = R^*$; the corresponding variables F_x and M_y can be calculated, and the results are in accordance with Goldman's results, as shown in Figures 4.10 and 4.11.

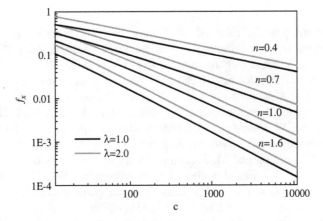

Figure 4.10: The dimensionless tangential resistance of interstitial fluid by equation (4.29).

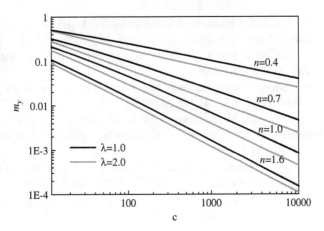

Figure 4.11: The dimensionless tangential resistance moment of interstitial fluid by equation (4.30).

In summary, wet granular material is composed of particles and the interstitial fluid, and forms very complex and variable configurations. It is not only different from dry granular material, but is also a liquid–solid two-phase flow. Besides the adhesion of the fluid between particles, the fluid itself moves when there is a relative movement between particles; this generates lubrication and viscosity. To summarize, it is very difficult to manage and understand such liquid–solid coupling in wet granular matter.

In the actual calculations, the liquid bridge force and the contact force between dry particles are usually analyzed separately. If there is no contact between particles, it is only necessary to consider the normal force, the tangential force, and the resistance moment generated by the interstitial fluid between particle gaps. If the particles have come into contact, the normal force and the tangential force can be obtained by using the contact models for dry particles, ignoring the liquid bridge force of the interstitial fluid.

References

[1] M. J. Adams and D. Edmondson 'Forces between particles in continuous and discrete liquid media', Tribology in Particulate Technology, (eds). B. J. Briscoe and M. J. Adams, Adam Hilger, Bristol, pp. 154–172 (1987).

[2] A. J. Goldman, R. G. Cox and H. Brenner, 'Slow viscous motion of a sphere parallel to a plane wall. II. Couette flow', Chem. Eng. Sci., 22, 653–660 (1967).

[3] S. Herminghaus, 'Dynamics of wet granular matter', Adv Phys., 54, 221-251 (2005).

[4] G. Lian, Y. Xu, W. Huang and M. J. Adams, 'On the squeeze flow of a power-law fluid between rigid spheres,' J. Non-Newtonian Fluid Mech., 100, 151164 (2001).

[5] G. Lian, C. Thornton and M. J. Adams, 'A theoretical study of liquid bridge forces between two rigid spherical bodies', J. Colloid Interface Sci., 161, 138–147 (1993).

[6] G. Lian, C. Thornton and M. J. Adams, 'Discrete particle simulation of agglomerate impact coalescence', Chem Eng. Sci., 53, 3381–3391 (1998).

[7] G. Lian, C. Thornton and M. J. Adams, 'A theoretical study of the liquid bridge forces between two rigid spherical bodies', J. Colloid and Interface Sci., 161, 138–147 (1993).

[8] N. Lu and W. J. Likos, Unsaturated Soil Mechanics, Wiley, New York (2004).

[9] N. Mitarai and F. Nori, 'Wet granular materials', Adv. Phys., 55(1–2), 1–45 (2006).

[10] O. Pitois, P. Moucheront and X. Chateau, 'Rupture energy of a pendular liquid bridge', Eur. Phys. J.B, 23, 79–86 (2001).

[11] G. J. Rodin, 'Squeeze film between two spheres in a power-law fluid', J. Non-Newtonian Fluid Mech., 63, 141–152 (1996).

[12] M. Scheel, R. Seemann, M. Brinkmann, M. Di Michiel, A. Sheppard, B. Breidenbach and S. Herminghaus, 'Morphological clues to wet granular pile stability', Nature Mater., 7(3), 189–193 (2008).

[13] C. D. Willett, M. J. Adams, S. A. Johnson and J. P. K. Seville, 'Capillary bridges between two spherical Bodies,' Langmuir, 16, 9396–9405 (2000).

Chapter 5

Discrete element method

In conventional methods, the granular material system is treated as a continuum by averaging the physical characteristics across many particles. In the case of solid-like granular behavior, as in soil mechanics, the continuum approach usually treats the material as elastic or elastoplastic and models it with the finite element method or a mesh-free method. In the case of liquid-like or gas-like granular flow, the continuum approach may treat the material as a fluid and use computational fluid dynamics. The drawbacks to homogenization of the granular scale physics, however, are well documented and should be carefully considered before any attempt to use a continuum approach.

The discrete element method (DEM) is an effective numerical method for computing the motion of a large number of macroscopic particles. The DEM is becoming widely accepted in cases in which the model system is composed of any discontinuous materials (including soils and rock fragments) rather than a continuum. The DEM includes explicit and time-matching algorithms to solve the equations of motion. Particles may be rigid or deformable, but contacts are always deformable. All these forces are summed to find the total force acting on each particle. An integration method is employed to calculate the change in the position and the velocity of each particle during a certain time step, using Newton's laws of motion. Then, the new positions are used to calculate the forces during the next step, and this loop is repeated until the simulation ends. In granular statics, the DEM calculates the equilibrium states of particle systems by using dynamics transitions, the convergence of which are generally accelerated and optimized by introducing an artificial viscous damping.

5.1 Contact searching

The simulation of granular systems requires cyclic calculations of particle displacement increments and contact force increments. For the calculation of contact forces among the particles, an efficient searching scheme is needed to detect the contacts among the particles. Because a particle cannot make contact with particles which are distant, the principle is to search for contacts with only those particles in the local vicinity.

In the DEM, the workspace is divided into many subcells and is so chosen that the size of the cell is larger than the diameter of the largest spheres. However, in order for the DEM to search the contact efficiently, the cell size l_{box} should normally be less than twice the largest sphere diameter d_{max}:

$$d_{max} < l_{box} < 2d_{max} \qquad (5.1)$$

The first step in the contact detection scheme is to form link lists for each cell by mapping all constituent particles into the appropriate cells. A spherical particle would be mapped into a cell if one or more corners of the circumscribing cube of dimension $(l_{box} + TOL)$ fall into the cell. Because the maximum radius of the constituent particles is restricted such that $d_{max} < (l_{box}/2 + TOL)$, the maximum number of cells in which a particle has entries is eight, and the maximum number of cells is four for the two-dimensional (2D) case.

The mapping scheme in case of 2D is illustrated in Figure 5.1. Particle A is located in cell 3, B in cells 5, 6, 8, and 9, C in cells 7 and 8, and D in cell 9. Hence, one particle is mapped to at least one cell, and at most to four cells. Before checking whether one particle is in contact with others, it is also necessary to check that particles are mapped into the same cell. Particles B and C possibly contact with each other because both mapped into cell 8, but it is impossible for particles A and C to be in contact. The second step is the search for contacts. Since possible contacts exist only between those particles which are mapped into the same cell, contact searching only needs to be performed between those spheres that are held in the link lists of each cell. With the DEM, the contact search is actually performed in the same process during the mapping for the link list forming. Each time that a sphere is mapped into a cell, the spheres that were previously mapped into the same cell are searched and their distances to the newly mapped sphere are calculated. Any sphere in the link list lying within a specified gap distance TOL of the newly mapped spheres is regard as a potential contact.

5.2 Rigid-sphere-based DEM

In the rigid-sphere model, there is no force acting on the particles except at the instant. Thus, the rigid-sphere model can be viewed as an approximation to the soft-sphere model with high-velocity particles. The dynamics of the particles consist of an alternating sequence of flight segments and instantaneous binary collisions. Particles travel in straight lines with

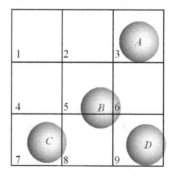

Figure 5.1: The mapping scheme for spheres.

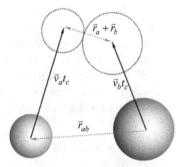

Figure 5.2: Free motion time of two rigid spheres.

constant speed between collisions. To the lowest order in the density, these multiple collisions and finite collision time effects may be ignored, and the dynamics may be thought of as periods of free motion of the particles punctuated by binary collisions between pairs of particles. During the free motion, the positions and velocities of the particles change with time. As shown in Figure 5.2, the free motion time is determined as follows:

$$t = \left\{ \vec{r}_{ab}\vec{v}_{ab} - \sqrt{(\vec{r}_{ab}\vec{v}_{ab})^2 - v_{ab}^2 \left[\vec{r}_{ab}^2 - (\vec{r}_a + \vec{r}_b)^2 \right]} \right\} / v_{ab}^2 \tag{5.2}$$

where $\vec{r}_{ab} \equiv \vec{r}_a - \vec{r}_b$, and $\vec{v}_{ab} \equiv \vec{v}_a - \vec{v}_b$. $\vec{r}_a$ and $\vec{r}_b$ are the positions of particles a and b, respectively; $\vec{v}_a$ and $\vec{v}_b$ are their velocities.

The condition for a collision is:

$$\begin{cases} \vec{r}_{ab}\vec{v}_{ab} < 0 \\ (\vec{r}_{ab}\vec{v}_{ab})^2 - v_{ab}^2 \left[\vec{r}_{ab}^2 - (\vec{r}_a + \vec{r}_b)^2 \right] \geq 0 \end{cases} \tag{5.3}$$

During an instantaneous collision between spheres, the post-collision positions and laboratory momentum are related to their pre-collision values. In order to follow the motions of the particles, we only need to know when the collisions occur and the velocities after collision.

5.3 Soft-sphere-based DEM

5.3.1 Dynamic relaxation

The dynamic relaxation method is based on discretizing the continuum under consideration, by summing the mass at nodes and defining the relationship between nodes in terms of stiffness (see also the finite element method). The system oscillates about the equilibrium position under the influence of loads. An iterative process is followed by simulating a pseudo-dynamic process in time, and all iterations are based on an update of the geometry. For granular materials, the equilibrium equation of the particle may be approached from the point of view of dynamic relaxation, by adding fictitious inertial and viscous forces. The dynamic approach of the statics of granular materials is especially suited for describing the possible free-fall motion of particles which momentarily lose contact with other particles. Rayleigh damping is widely used for dynamics relaxations, which involves mass-proportional (global) damping and stiffness-proportional (contact) damping. Note that contact damping may require reduction in time steps for numerical stability.

Classical Rayleigh damping is based on a damping matrix C that is proportional to the mass and stiffness matrices as follows:

$$C = \alpha M + \beta K \tag{5.4}$$

where α is the mass proportional Rayleigh damping coefficient, β is the stiffness proportional Rayleigh damping coefficient, M is the system structural mass matrix, and K is the system structural stiffness matrix.

With this formulation, the damping ratio is the same for axial, bending, and torsional responses. Classical Rayleigh damping results in different damping ratios for different response frequencies according to the following equation:

$$\xi = \frac{1}{2}\left(\frac{\alpha}{\omega} + \beta\omega\right) \tag{5.5}$$

where ξ is the damping ratio (a value of 1 corresponds to critical damping), and ω is the response frequency in rad/s. It can be seen from this that the mass proportional term produces a damping ratio inversely proportional to the response frequency, and the stiffness proportional term produces a damping ratio linearly proportional to the response frequency. In many situations for granular materials, we empirically choose $\omega = \pi$, $\xi \approx 0.05 \sim 0.1$, and then calculate α and β as follows:

$$\text{As } \beta = 0, \ \alpha = 2\xi\omega \tag{5.6}$$

$$\text{as } \alpha = 0, \ \beta = \frac{2\xi}{\omega} \tag{5.7}$$

5.3.2 Numerical scheme

Typical integration methods used in a DEM are the Euler method and the Verlet algorithm.

5.3.2.1 Euler method
The Euler method is a first-order numerical procedure for solving ordinary differential equations with a given initial value. It is the most basic kind of explicit method for numerical integration of ordinary differential equations.

At time t, the position of particles is $r_i(t)$. For a small time step Δt, usage of the first two terms of the Taylor expansion of y represents the linear approximation around the point $r_i(t)$. One step of the Euler method from t to $t + \Delta t$ is:

$$r_i(t + \Delta t) \approx r_i(t) + v_i(t)\Delta t \tag{5.8}$$

The Euler method is explicit, that is, the solution of $r_i(t + \Delta t)$ is an explicit function of $r_i(t)$. The velocity of particle i is represented through a similar expression:

$$v_i(t + \Delta t) \approx v_i(t) + \frac{1}{m_i} F_i(r_i(t))\Delta t \tag{5.9}$$

It can be seen that when initial values are given, the positions and velocities of particles at the next step can be explicitly obtained. Coding is relatively simple and the program runs quickly.

The magnitude of the errors arising from the Euler method can be demonstrated by comparison with a Taylor expansion. For small Δt, the dominant error per step, or the local truncation error, is proportional to $(\Delta t)^2$. To solve the problem over a given range of t, the number of steps needed is proportional to $1/\Delta t$, so it is to be expected that the total error at the end of the fixed time, or the global truncation error, will be proportional to h (error per step multiplied by the number of steps). Because the global truncation error is proportional to h, the Euler method is said to be of first order. This makes the Euler method less accurate (for small h) than other higher order techniques such as Runge–Kutta methods and linear multistep methods.

5.3.2.2 Verlet integration
Verlet integration is a numerical method used to integrate Newton's equations of motion. It is frequently used to calculate trajectories of particles in molecular dynamics simulations. The Verlet integrator offers greater stability than the much simpler Euler method, as well as considers other properties that are important in physical systems, such as time-reversibility and area-preserving properties. At first, it may seem natural to simply calculate trajectories using Euler integration. However, this type of integration suffers from many problems, as discussed earlier. The stability of the technique depends quite heavily on either a uniform update rate, or the ability to accurately identify positions at a small time interval (Δt) into the past.

Verlet integration is the simplest second-order method. The position and velocity at $t + \Delta t$ are expressed by:

$$r_i(t + \Delta t) \approx 2r_i(t) - r_i(t - \Delta t) + \frac{1}{m_i} F_i(r_i(t))\Delta t^2 \tag{5.10}$$

$$v_i(t) \approx \frac{1}{2\Delta t}[r_i(t + \Delta t) - r_i(t - \Delta t)] \tag{5.11}$$

In comparison with equations (5.8) and (5.9), Verlet integration is a more accurate higher order method than the Euler method.

The local error in position of the Verlet integrator is $O(\Delta t^4)$ as described earlier, and the local error in velocity is $O(\Delta t^2)$. The global error in position, in contrast, is $O(\Delta t^2)$ and the global error in velocity is $O(\Delta t^2)$.

5.3.3 System evolution

By using the Euler method to simulate the evolutions of contact forces and particle movements, the forces, and hence the accelerations, are assumed to be constant over the interval of each time step. With global damping, the Newton's second law of motion for each sphere is given as:

$$F_i(t) + mg_i - \beta_g v_i(t) = m\frac{\Delta v_i}{\Delta t} \tag{5.12}$$

$$M_i - \beta_g \omega_i(t) = I\frac{\Delta \omega_i}{\Delta t} \tag{5.13}$$

From this, the updated velocity components are given as:

$$v_i(t) = v_i(t - \Delta t)\frac{m/\Delta t - \beta_g/2}{m/\Delta t + \beta_g/2} + \frac{F_i(t) + mg_i}{m/\Delta t + \beta_g/2} \tag{5.14}$$

$$\omega_i(t) = \omega_i(t - \Delta t)\frac{I/\Delta t - \beta_g/2}{I/\Delta t - \beta_g/2} + \frac{M_i(t)}{I/\Delta t - \beta_g/2} \tag{5.15}$$

where $i = 1, 2$, and 3 indicates the three components in x, y, and z directions. β_g is the global damping coefficient and Δt is the time step. F_i represents the components of the out-of-balance force; M_i represents the components of the out-of-balance momentum; g_i is the acceleration due to gravity; m is the mass of the sphere; I is the rotational inertia of the sphere; $v_i(t)$ and $\omega_i(t)$ are the updated components of the translational and rotational velocities; and $v_i(t - \Delta t)$ and $\omega_i(t - \Delta t)$ are the previous components.

From the updated velocity components of each particle, the calculation of the components of the translational and rotational displacements, as increments in the time step, is according to the following equations:

$$\Delta x(t) = v_i(t)\Delta t \tag{5.16}$$

$$\Delta \theta(t) = \omega_i(t)\Delta t \tag{5.17}$$

The translational displacement increment is added to the accumulated displacement increment.

5.3.4 Time step

DEMs apply explicit time integration. Explicit time integration requires the time step to be smaller than a critical value to maintain the stability of the solution. In finite element simulations, explicit solution schemes are well known. The stability and the accuracy of the numerical solution are influenced by the Courant criterion. The Courant criterion couples the element size and the critical time step as follows:

$$\Delta t \leq \Delta x/c \tag{5.18}$$

where Δx is the grid size and c is the sound speed. An excessively large time step causes instability. In the DEM, the evolution of a system is based on the collisions of the particles. It is reasonable to assume that there may be geometrical and physical limitations on the time step.

Particles i and j are not allowed to penetrate each other along their central line at one time step. This means:

$$\Delta t \le \Delta t_{cr}^{G} = \left(r_0^i + r_0^j\right)/v_n^{ij} \tag{5.19}$$

where Δt_{cr}^{G} is the critical time step based on the geometric limitation, and v_n^{ij} is the central relative speed at time step n.

For a binary collision, the momentum transferred from one particle to another in a single time step should not exceed the total momentum exchange during the collision process. Assuming that particles i and j come into contact at time t_0, the translational velocity of particle i is v_n^i, and the translational velocity of particle j is zero, and then, at $t0 + \Delta t$, the following equation applies:

$$\Delta u^{ij} = v_n^i \Delta t, \, \varepsilon^{ij} = \Delta u^{ij}/r_0^{ij}, \, F^{ij} = F\left(\varepsilon^{ij}\right)E_0^{ij}A_0^{ij}\varepsilon^{ij} \tag{5.20}$$

Here, Δu^{ij} is the increment of the relative displacement; ε^{ij} is the strain; A_0^{ij} is the average cross-sectional area of particles i and j; and E_0^{ij} is the equivalent modulus, as shown in the equation: $E_0^{ij} = 2E_0^i E_0^j/\left(E_0^i + E_0^j\right)$. E_0^i and E_0^j are the modulii of particles i and j.

The physical restriction can be written as:

$$F^{ij}\Delta t = m^j v^j \le M \tag{5.21}$$

where $F^{ij}\Delta t$ is the impulse in one time step, $m^i v^j$ is the momentum the particle j obtained, and M is the maximum momentum the particle j can gain during the collision. According to classic mechanics, M is written as:

$$M = \frac{(2-e)m^i m^j}{m^i + m^j}v_n^{ij} \tag{5.22}$$

where e is the restitution coefficient and v_n^{ij} is the axial relative velocity of two particles before the collision. Substituting equations (5.20) and (5.22) into equation (5.21) yields the following:

$$\Delta t \le \Delta t_{cr}^{p} = \sqrt{\frac{(2-e)m^i r_0^{ij}}{(1+\beta)E_0^{ij}A_0^{ij}}} \tag{5.23}$$

where Δt_{cr}^{p} is the physically critical time step and $\beta = m^i/m^j$. Assuming that $r_0^{ij} = r_0^i + r_0^j = 2r_0^i = d_0^i$, equation (5.23) turns out to be as follows:

$$\Delta t \le \frac{d_0^i}{C_{ij}'} \tag{5.24}$$

Here, C_{ij}' is the equivalent sound speed.

$$C_{ij}' = \sqrt{\frac{3(1+\beta)E_0^{ij}}{2(2-e)\rho^i}} \tag{5.25}$$

The physical meaning of equation (5.24) is obvious: the propagation of equivalent sound distribution should not exceed the size of a particle in one time step, otherwise it may cause nonphysical results. Hence, this requirement is actually equivalent to the Courant stability condition. Equation (5.25) is applicable toward the linear relationship of F^{ij} with the strain ε^{ij}. Corrections should be made for nonlinear cases, since C_{ij}' changes with ε^{ij}.

Δt_{cr}^G and Δt_{cr}^P, as discussed earlier, are obtained by assuming that the collision is over in one time step. To increase the time resolution of the collision process, the number of the time steps should be more than one, hence the critical time step in the practical calculation should be:

$$\Delta t_{cr} = \frac{1}{k} \min \left\{ \Delta t_{cr}^G, \Delta t_{cr}^P \right\} \tag{5.26}$$

Here, k is the number of time steps.

Another method to determine the critical time step is to consider the Rayleigh wave propagation. A Rayleigh wave is a type of surface acoustic wave that travels on solids, across their surfaces. In isotropic solids, the surface particles move in ellipses in planes normal to the surface and parallel to the direction of propagation. At the surface and at shallow depths, this motion is retrograde. Particles deeper in the material move in smaller ellipses with an eccentricity that changes with depth. At larger depths, the particle motion becomes prograde. The depth of significant displacement in the solid is approximately equal to the acoustic wavelength. It was found that approximately 70% of the dissipated energy is due to Rayleigh wave propagation. Upon the application of a force on an elastic body, the Rayleigh waves are propagated along the surface with a velocity v_R:

$$v_R = \beta \sqrt{\frac{G}{\rho}} \tag{5.27}$$

where G and ρ are the shear modulus and density of the materials; β is the root of the following equation:

$$(2 - \beta^2)^4 = 16(1 - \beta^2)\left[1 - \frac{1 - 2v}{2(1 - v)}\beta^2\right] \tag{5.28}$$

Here, v is the Poisson's ratio. From this, an approximate solution may be obtained as:

$$\beta = 0.163v + 0.877 \tag{5.29}$$

Equation (5.27) could be rewritten as:

$$v_R = (0.163v + 0.877)\sqrt{\frac{G}{\rho}} \tag{5.30}$$

For an assembly of many spherical particles, it can be derived that the highest frequency of Rayleigh wave propagation is determined by the smallest spheres, which in turn gives the critical time step as:

$$\Delta t = \frac{\pi R}{v_R} = \frac{\pi R}{0.163v + 0.877}\sqrt{\frac{\rho}{G}} \tag{5.31}$$

Here it is assumed that the property type of all constituent particles is the same. However, if there are different material types for the constituent particles, the critical time step for the highest Rayleigh wave frequency should be the lowest among those determined by different material types, as follows.

$$\Delta t = \pi \left[\frac{R}{0.163\nu + 0.877} \sqrt{\frac{\rho}{G}} \right]_{min} \tag{5.32}$$

The actual time step is a multiple of the Rayleigh critical time step by a value which is normally given as less than 1. It has been shown that, for most cases, the time step based on the Rayleigh wave speed can ensure the numerical stability of simulations. However, for the simulation of some dynamics processes, the Rayleigh time step may not be small enough to guarantee numerical stability. This is because, for the earlier Rayleigh wave propagation based time step, the relative motions between particles are not considered. Since the relative velocity between particles may be very high, the Rayleigh wave transmission through the assembly along each sphere surface will be greatly affected. As a result, the critical time step of real Rayleigh wave transmission is much smaller and the Rayleigh wave time step based on the static assembly does not apply to the dynamic assembly any more. When numerical instability occurs during the simulations of dynamics processes of a granular assembly, the time step needs to be further decreased.

5.4 Large-scale parallel computing

DEMs are relatively computationally intensive, which limit either the length of a simulation or the number of particles. Several DEM codes take advantage of parallel processing capabilities (shared or distributed systems) to scale up the number of particles or length of the simulation.

In computational mathematics, the degree of confidence is used to estimate the similarity between a scientific computation and objective physical phenomenon. It includes the degree of confidence for physical modeling and the degree of confidence for numerical computation. Physical modeling is used to bring up mathematical-physical equations and provide physics parameters according to the operational pattern and evolution of the objective system. Numerical computation is employed to enable discretization of the equations, design appropriate computing methods, and develop the program, combined with the geometrical configuration and the physical parameters of the physical problem. To obtain the numerical solutions, the program is run on computers. If the mathematical-physical equations cannot reflect the primary characteristics of the objective system, or if the computing method is not suitable for the mathematical-physical equations, the numerical solution may lose its degree of confidence as well as its scientific value.

In order to achieve accurate and scientific physical modeling, the research objectives must focus on a large-scale system, rather than on a small-scale system such as the one under study. The process may be transformed from the quasi-static status to the dynamic status, and also from the simplified 2D to the realistic 3D. All the above factors cause the computation scale to surge by a magnitude of several thousand, and therefore this places extremely urgent demands on the efficient usage of computer resources. For example, when considering the contact mechanics of particles, the contact law is usually simplified into the soft-sphere model. However, the high-confidence computing requires a rigorous Hertz–Mindlin contact model and Thornton model. With the inclusion of additional considerations on particle rolling, the model will become more complicated. With current computation capabilities, the simulation scale in 3D finds it difficult to achieve the level of 10^5 particles. In other words, if a particle is 0.5 mm in radius, with about 50 such particles located in each direction, we can only simulate a cube measuring 2.5 cm in side length. This computational scale is far from the laboratory scale, and thus no reasonable comparison could be established between the numerical simulation and the experiment. If we want to simulate a cube of 10 cm in side length, about 6 million particles are required. Further, when including other considerations such as the speed of iteration convergence, the real computation scale will be far greater than the simplified estimation presented earlier. To successfully face this severe challenge, large-scale parallel computation must be developed.

In the past decade, the peak performance of parallel computers in our country has surged by a multiple of 1,000, and has currently reached 4.7 trillion cyclings per second. Meanwhile, the CPU cores of parallel computers have begun numbering in the hundreds of thousands; the systems are becoming increasingly complex. Because physical modeling tends to be finer, and the computer system architecture becomes more complicated, the research and development of application programs are more difficult and the development period is longer. The currently used parallel particle application programs are mostly obtained through the parallelization of serial programs. As a result, the computational efficiency is relatively low; this cannot satisfy the needs of scientific research in terms of the rapid development of computers. In order to solve these difficulties, we should modernize our ideas in developing application programs and insist on multidisciplinary integration, such as physics, mechanics, computational mathematics, and computer science. The physicists and mechanical scientists are responsible for the physical modeling, while the computational mathematicians and computer scientists are responsible for the computational method and the development of DEM application programs. Based on the characteristics of granular systems, computer scientists are responsible for improving the single-core float performance and the multicore parallel efficiency in the DEM application programs.

Currently, Tsinghua University in Beijing has initiated the development of DEM parallel programs based on the J Adaptive Structured Mesh applications Infrastructure (JASMIN) developed by the Institute of Applied Physics and Computational Mathematics of Beijing. A simple simulation is presented in Figure 5.3.

Our aim is to establish a large-scale parallel computation platform for granular materials based on the super computers and JASMIN and build a solid foundation for the development of granular material scientific computational capability with a high degree of confidence. Based on this platform, we intend to complete the simulations for the typical mechanical behavior of large-scaled systems (with a magnitude of 10^6 to 10^7 particles), analyze their internal structures and evolution laws, and discover the physical structure behind the complex macroscopic mechanical behavior. Meanwhile, in relation with some important mechanical characteristics of engineering practices, we intend to carry out large-scale computations and analysis.

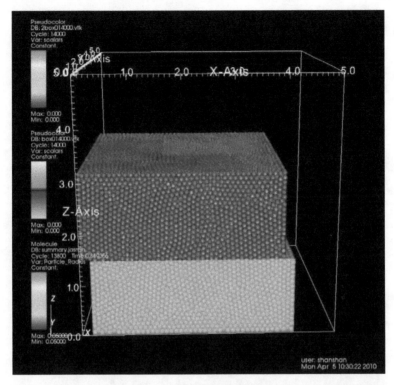

Figure 5.3: Direct shear simulations of 140,000 particles with eight cores.

References

[1] P. A. Cundall and O. D. L. Strack, 'A discrete numerical model for granular assemblies', *Geotechnique*, 29(1), 47–65 (1979).

[2] T. Pöschel and T. Schwager, *Computational Granular Dynamics: Models and Algorithms*, Springer-Verlag, Heidelberg, Germany (2005).

[3] D. C. Rapaport, *The Art of Molecular Dynamics Simulation*, Cambridge University Press, Cambridge, UK (2004).

[4] C. Thornton (ed.), 'Numerical simulations of discrete particle systems', *Powder Technol.*, 109 (Special Issue), 29301 (2000).

Chapter 6

Force chains

In dense granular systems, the transmission of external loadings from one boundary to another can only occur via interparticle contacts, and a multiplicity of pathways may be established in order to achieve a stable stress state. Some particles with larger contact deformation and quasi-linear mechanisms of contact transmit a large fraction of the loading forces, which forms the strong force chains; the other particles exist with smaller contact deformation and transmit a small loading force, thereby forming the weak force chains. Although strong force chains are not particularly numerous, they dominate the unique phenomena and mechanical properties of granular matter. For example, a small shock in the strong force chains in piles of sand may lead to massive avalanches. A larger number of weak force chains are approximately evenly distributed within the granular matter, thereby supporting the strong force chains.

6.1 Formation of force chain

In dense packings, the space for free movement of particles is small; therefore, there is an extrusion deformation between particles, induced by external loadings. Normally, the interparticle forces in the same force chain are almost equal. The direction of a force chain is almost parallel to the major principal stress in the granular assembly. In weak force chains, the particles experience a slight deformation, and thus, a small tangential force can break the weak force chains, as shown in Figure 6.1(a). In strong force chains, the particles bear a large part of the external loadings and have a relatively large contact area. When the contact force between particles is within the frictional cone (as shown by the dashed line in Figure 6.1(b)), the particles stay in a self-locking state, and therefore, the strong force chains can afford a greater tangential force. Obviously, the larger the particle surface friction coefficient is, the greater the tangential force borne by the force chain is. One extreme in this scenario is when the particle surface is smooth; in this case, force chains cannot tolerate any additional tangential force and the granular assembly becomes more fragile.

Since a particle can maintain contact with only a limited number of neighbors (i.e., the number is the so-called coordination number), the anisotropy of force chains is very obvious, as illustrated in Figure 6.2. For monodispersed particles in two dimensional, the maximum

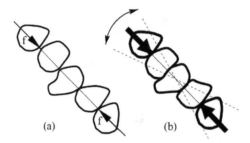

Figure 6.1: Particle extrusion deformation in a force chain.

coordination number is 6. The gray lines represent weak force chains where the contact forces are less than the mean contact force, while the black lines represent the strong force chains where the contact forces are greater than the mean contact force. The line thickness is proportional to the magnitude of contact forces. It can be seen that although there are few strong force chains, the weak force chains spread around the strong force chains, thereby providing a strong support structure to maintain the stability of the former.

With regard to the morphology of force chains, Horne put forward the early concept of "*solid paths*" in 1965; Edwards and Oakeshott proposed the arching concept of granular matter in 1989, which was deepened by Bouchaud et al. and Cates et al., in 1995. They clearly put forward the concept of force chains "... *Arches are chain-like configurations of grains..., which act to transport force along the chains...*" A lot of experimental measurements have been performed and have verified the existence of force chains, for example, by detecting the contact force distribution of certain sections of granular matter. These measurements have used methods such as the high-precision electronic balance weighing method, the carbon paper indentation method, and photoelastic stress analysis, among others. However, the main drawback of the experimental methods is that the weak

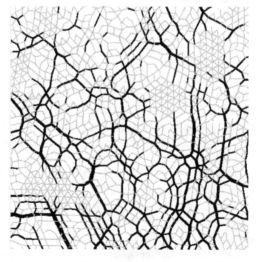

Figure 6.2: The force chain network of a statically packed two-dimensional granular assembly.

forces cannot be accurately measured, and the force chain structure cannot be invasively detected. Therefore, it is necessary to conduct numerical simulations based on rigorous contact mechanics. The effects of material properties, initial conditions, and boundary conditions to contact force distribution and force chain networks can be completely analyzed.

6.2 Measurements of contact force

6.2.1 Photoelastic stress analysis

Photoelasticity is an experimental method to determine the stress distribution in a material. The method is based on the property of birefringence in certain transparent materials. Photoelastic materials exhibit the property of birefringence only on the application of stress; the magnitude of the refractive indices at each point in the material is directly related to the state of stress at that point.

Since the 1960s, the development of laser technology has provided a high-intensity light source, which has promoted the wide use of photoelasticity for a variety of stress analyses. This was particularly the case before the advent of numerical methods, such as finite elements or boundary elements. The applications of photoelasticity in investigating the highly localized stress state within granular assemblies started in around the 1970s, and now it is a basic method for observing force chain structure and for determining interparticle forces.

The basic device in the photoelastic experiments is the plane-polarized light device, which is composed of a light source and a pair of polarizers; the one near the light source is called the close polarizer, and the other is called the analyzer, as shown in Figure 6.3. When the two polarizers are on an orthogonal axis, there forms a dark field; usually, a polarizer is adjusted to be on the vertical axis, while the other is horizontal. When the axes of the two polarizers are parallel to each other, there forms a light field.

In the orthogonal plane-polarized light field, a force that is exerted on the model made from birefringent materials gets resolved along the two principal stress directions, and each of these components experiences different refractive indices. The difference in the refractive indices leads to a relative phase retardation between the two component waves. The relative retardation Δ is given by the stress optic law:

$$\Delta = Ch(\sigma_1 - \sigma_2) \tag{6.1}$$

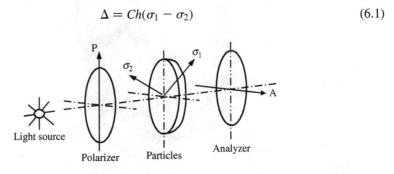

Figure 6.3: The plane polarized light device.

where C is the stress optic coefficient, h is the specimen thickness, σ_1 is the first principal stress, and σ_2 is the second principal stress

When Δ is an integer multiple of wavelength λ, then

$$\Delta = N\lambda \tag{6.2}$$

Extinction interference occurs and displays some dark points in the specimen, and many such points form a black line. This phenomenon is known as isochromatics, which can be obtained by equations (6.1) and (6.2)

$$\sigma_1 - \sigma_2 = \frac{Nf}{h} \tag{6.3}$$

Here, $f = \frac{\lambda}{C}$ is the so-called material fringe value.

In a plane polariscope setup, the fringe pattern consists of both the isochromatics and the isoclinics. The isoclinics change with the orientation of the polariscope, while there is no change in the isochromatics. In order to eliminate the isoclinics and to obtain a clear isochromatic graph, a pair of 1/4 wave plates is placed between the two polarizers to form an orthogonal circularly polarized light field. When measuring stress in photoelastic particles, the typical material used is polycarbonate, which has high optical sensitivity, transparency, and a small creeping effect at room temperature. As shown in Figure 6.4, strong force chains are observed as being heterogeneously distributed throughout the assembly. Basically, brighter particles have higher internal stress. Figure 6.4(b) exhibits the fringe pattern inside a single particle, after which the normal force and tangential force at each contact can be resolved numerically. The corresponding distribution is shown in Figure 6.5.

In Figure 6.5, the horizontal coordinate is the ratio f of the normal force and the tangential force normalized with the mean normal force $\langle F_n \rangle$, and the vertical coordinate is the probability density distribution. It is found that after the normal force exceeds $\langle F_n \rangle$, it decreases exponentially, and that there is a peak at $\langle F_n \rangle$, namely, $f = 1$ in the figure.

Figure 6.4: Measurement of the interparticle forces according to the fringe pattern.

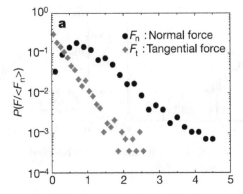

Figure 6.5: The normal force and the tangential force measured by the photoelastic method (Majmudar and Behringer, 2005).

6.2.2 Carbon paper method

A piece of carbon paper is placed between the granular assembly and a blank sheet. As the particles compress on the carbon paper, the pressure would deposit ink on the blank sheet. According to the sizes of ink on the blank paper, normal forces can be determined. Since the indentation color caused by the weight of particles is lighter, external loading is usually required. For example, Erikson et al. [7] selected soft rubber balls, whose diameter was 3.12 ± 0.05 mm, stacked in a cylinder with diameter 140 mm; the stacking height was 70 mm; finally, a 2500 to 7000 N force was exerted on the top of the cylinder, as illustrated in Figure 6.6. It was found that when the contact deformation is small, the distribution peak at $\langle F_n \rangle$ is not obvious; the normal force greater than $\langle F_n \rangle$ decreases exponentially and the trend is gentle. When the deformation is larger, for example, when the strain is 37%, the

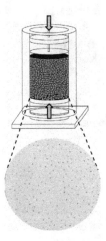

Figure 6.6: Measurement of the normal force of particles according to the sizes of ink impressions left on the blank paper (Erikson et al., 2002).

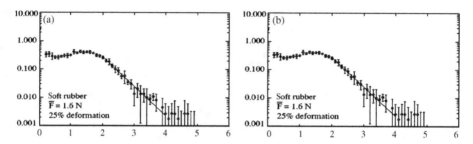

Figure 6.7: The distribution of normal forces of soft rubber balls. (a) The strain is 25%; (b) the strain is 37%.

distribution peak at $\langle F_n \rangle$ becomes increasingly obvious. The normal force greater than $\langle F_n \rangle$ not only decreases exponentially, but also decreases very rapidly. This can be seen in Figure 6.7(b).

6.2.3 Electronic balance weighing method

By using a high-precision electronic balance, Løvoll et al. (1999) detected the contact forces on bottom particles when the particles were statically packed under gravity, as illustrated in Figure 6.8. The particle diameter was 2 ± 0.05 mm, the cylinder diameter was 80 mm, and the cylinders were filled with approximately 120,000 particles weighing about 1.3 kg in total. The Mettler PM1200 electronic balance can weigh up to 1200 g, and the minimum indication is 1 µg, while the particle weight was 15.3 µg.

The detection probe of the electronic balance was approximately equal to the radius of a single particle. When sliding on the panel surface, the probe could exactly distinguish among

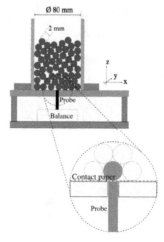

Figure 6.8: Measurement of contact forces on the bottom particles using the electronic balance (Løvoll et al. [12]).

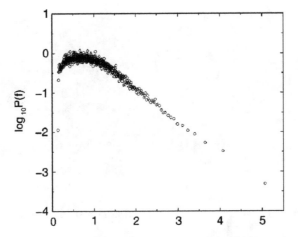

Figure 6.9: The force distribution as measured by an electronic balance.

particles and measure the contact force. However, the relative sliding of the probe against the particles may have affected the internal force chains, which could drive the contact forces belonging to different configurations of force networks.

The contact force distribution measured by the electronic balance is similar to those shown in Figures 6.5 and 6.7; it can be seen that the probability distribution of the average normal contact force achieves its peak and then decreases exponentially with the increase in contact force.

More advanced equipment have been used to measure the distribution of contact forces, such as by Zhou et al. (2006), who measured the contact force distribution among smooth microdroplets on a three-dimensional surface using a high-resolution laser confocal-focusing microscope. The average diameter of the droplet was $30\,\mu\text{m}$. The particle surfaces were coated with a monolayer of trioctylphosphine oxide (TOPO). The thickness of this layer was $4\,\text{nm}$. When the two droplets were squeezed, the fluorescent powder on the contact surface A_{ij} shone, and A_{ij} could be measured by the laser confocal-focusing microscope. Subsequently, the contact force between two particles is given by $f_{ij} = A_{ij}\gamma(R_i + R_j)/(R_iR_j)$, where γ is the surface tension and $R_{i,j}$ is the droplet radius. Such a result of force distribution in a three-dimensional granular system is reported as being similar to two-dimensional cases.

6.2.4 Discrete element method

The discrete element method (DEM) simulates the mechanical behavior of granular assemblies consisting of spherical particles. This is a time-dependent finite difference scheme and is used for various simulations. The DEM, which is based on rigorous contact mechanics, can compute contact forces and force chains in three-dimensional granular matter. The packing of 11,000 equal-sized spherical particles with a radius $r = 0.5\,\text{mm}$ is simulated. The material properties of particles and walls were assumed similar to sand, $\rho = 2650\,\text{kg/m}^3$, $E = 100\,\text{MPa}$, and $v = 0.3$. The μ_s was set as 0.2 or 0.4 in different samples.

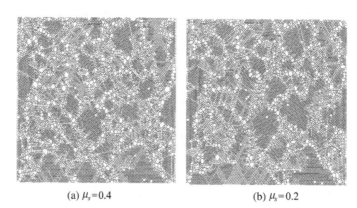

(a) $\mu_s = 0.4$ (b) $\mu_s = 0.2$

Figure 6.10: Contact networks at a packing fraction $\phi = 0.867$. (a) $\mu_s = 0.4$. (b) $\mu_s = 0.2$.

The contact networks are shown in Figure 6.10(a) and (b) for $\mu_s = 0.4$ and 0.2, respectively. The closely packed granular system is separated into many latticed blocks due to the monodispersion of the particles. At $\mu_s = 0.4$, there are a greater number of small blocks than at $\mu_s = 0.2$, and thus, the system appears highly disordered. Smaller values of μ_s easily induce crystallized or partially crystallized packings for monodispersed systems, while greater values of μ_s may increase the disorder.

The energy caused by deformation around the contact point is rather complicate to calculate. In this work, we simply consider the elastic energy due to normal compression, and analyze its distribution. For two smooth spheres with a contact radius a, the elastic energy is given by

$$W \approx \frac{1}{5R^*}\left(\frac{3R^*}{2E^*}(1-\nu^2)\right)^{2/3} F^{5/3} \quad (a/R^* \to 0). \tag{6.4}$$

As indicated in Figure 6.11, we found that around 60% of contacts carry forces less than $\langle F_n \rangle$, and the remaining 40% of contacts carry more than $\langle F_n \rangle$. We further note that these 40% of the contacts hold 80% of the energy in the system, which implies that those force chains carrying forces larger than the mean dominate the properties of granular matter. We propose that $F_n > \langle F_n \rangle$ is one of the conditions defining a force chain. It was shown that the shear stress is mainly determined by the strong network satisfying $F_n > \langle F_n \rangle$, while the strong and the "weak" networks (with $F_n < \langle F_n \rangle$) both contribute to the mean pressure.

Figure 6.12 indicates the probability density distributions of contact forces normalized by the corresponding forces. The distribution of the normalized interparticle force ($F_T/<F_T>$) is similar to that of the normalized normal forces ($F_n/\langle F_n \rangle$) shown as the inset in Figure 6.12(a), where F_T and $<F_T>$ are the interparticle force and the averaged interparticle force, respectively; F_n and $\langle F_n \rangle$ are the normal force and the mean normal force, respectively. Both have an exponential tail for a greater force and a dip toward zero for a smaller force. The tangential force F_t is one order smaller than F_n. At $\mu_s = 0.2$, $F_t = 0.0015$ N, and $F_n = 0.0183$ N. The distribution of the normalized tangential force $F_t/<F_t>$ is exponentially reduced as $F_t/<F_t>$ increases, as shown in Figure 6.12(b). From Figure 6.12(a) and (b), we

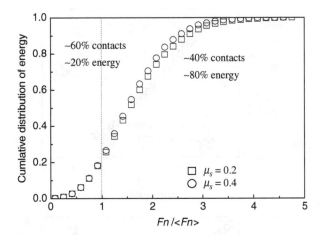

Figure 6.11: Distribution of elastic energy due to normal compression around the contact points. Around 40% of the contacts carry more than the mean force and hold 80% of the energy in the system, at $\mu_s = 0$.

can see that their shape is unaffected by μ_s. Figure 6.12(c) is the distribution of the ratio of $F_t/(F_n\mu_s)$. We find that at most contacts, $F_t/(F_n\mu_s)$ is less than 1. At larger μ_s, higher probabilities are located at smaller $F_t/(F_n\mu_s)$. The inset depicts the variation of $F_t/(F_n\mu_s)$ with $F_T/<F_T>$, and we cannot find a direct correlation between them.

The above measurement methods are summarized in Table 6.1, including the basic principles and advantages. Recently, more advanced methods have been borrowed from medical and biotechnological aspects for measuring the interparticle forces and distinguishing the spatial structures of force networks.

6.3 Bulk contact stress

When the packing fraction is small, particles distribute loosely in granular systems. When the packing fraction is greater than a threshold value, these particles compress one another and form force chains to support external loadings. The direction of force chains is approximately parallel to the direction of the principal stress. The unique features of granular material arise from the manner in which force is internally transmitted. In continuum mechanics, this is represented by a contact stress. In other words, a force F applied at a contact can be thought of as being transmitted in the direction of the vector l that connects the centers of mass of the two particles involved (the length of l is the distance between the particle centers). When averaging $\langle \rangle$ over time and volume, this yields the contact stress tensor ratio,

$$\frac{\tau_{xy}}{\tau_{yy}} = \frac{\langle F_x l_y \rangle}{\langle F_y l_y \rangle}$$

(6.5)

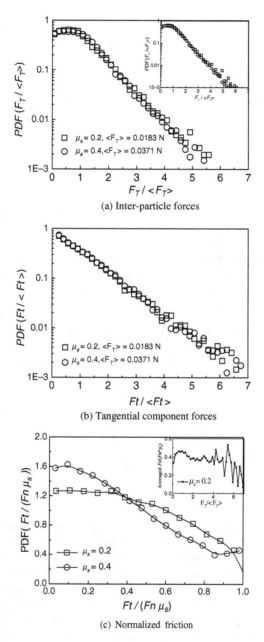

(a) Inter-particle forces

(b) Tangential component forces

(c) Normalized friction

Figure 6.12: Probability distributions of forces at $\phi = 0.867$. (a) The interparticle forces F_T. (b) The tangential forces F_t. (c) The normalized friction $F_t/(\mu_s F_n)$. The inset in (a) is the distribution of the normal forces. The inset in (c) is the variation of $F_t/(\mu_s F_n)$ with $F_T/<F_T>$.

Table 6.1: Comparison of measurement methods of the interparticle forces

Measurement methods	Representative work	Basic principle	Advantages
Carbon paper method	Erikson et al.	Force is determined with the ink impression size	Simple to use, but large loading is required to produce a measurable normal force
Weighing method	Løvoll et al.	Direct measurement of forces with an electronic balance	Small contact forces could be measured
Photoelasticity	Majmudar and Behringer	Calculation of stress distribution according to the fringe pattern, and generation of a fairly accurate picture of stress distribution	The normal and tangential forces can be measured simultaneously
Discrete element method	Sun and Wang	Based on rigorous contact mechanics	Three-dimensional contacts

Here, the components of force F in x- and y-directions are F_x and F_y, respectively; l_x and l_y are the components of distance vector l between two particles in the x- and y-directions.

When the force chains are extruded or sheared, the force chains themselves deform and generate resistance, and the frictional behavior of dense granular materials is a reflection of the evolution of the force chain structure. When ignoring the particle surface adhesion, the stress ratio τ_{xy}/τ_{yy} (i.e., the apparent friction coefficient $\tan\phi$, where ϕ is the internal friction angle) only depends on the geometry of force chains. As force chains form in the direction best suited to resist the applied forces, and as they collapse before they have rotated to any significant degree, their geometry is roughly fixed and controlled by the applied force. Therefore, the response of a dense granular assembly is the result of the force chain structure resisting the external loading, rather than the result of surface friction between the particles, as is shown in Figure 6.13.

Granular materials are often soft in bulk. This is evidenced in their sound speed, which is of the order of 100 m/s, roughly 50 times slower than the speed of sound in their constituent solid material, indicating that the bulk granular material has an apparent elastic modulus more than three orders of magnitude smaller than in its constituent solid. The bulk elastic modulus E_{bulk} of a random granular material from the contact stiffness can be shown as follows:

$$N \propto E_{\text{bulk}}(R^*)^{1/2}\alpha^{3/2} \sim E_{\text{bulk}}R^{1/2}\alpha^{3/2} \tag{6.6}$$

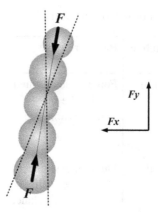

Figure 6.13: Contact stress and bulk frictions are related to force chain network.

While $N \propto k\alpha$, then

$$E_{bulk} \sim \frac{k}{\sqrt{R\alpha}} \sim \frac{kf(N)}{R} \tag{6.7}$$

Here, n is the coordination number of particles; the larger n is, the higher the ratio of the normal force against external force, leading to a greater strength of granular matter. The particle elastic coefficient k reflects the contact mechanical properties of particles and affects the overall elastic modulus of granular matter, and therefore, has a greater impact on the rheological properties of dense granular flow. When the packing fraction is larger, particles are fixed on force chains. For a given packing fraction, the deformation of a force chain (i.e., the sum of each pair of contact deformation between particles) is determined by the external loading. In equation (6.7), the elastic model of granular matter E_{bulk} is linear not only with k, but is also related to the particle radius R (i.e., the characteristic geometric length of the force chain).

The sound speed in constituent particles depends on the material properties of $\sim (E/\rho)^{1/2}$, and the sound speed in a static granular matter system is $\sim (E_{bulk}/\rho_{bulk})^{1/2} \sim k^{1/2}$, which can be used to detect the elastic coefficient of granular matter. Assuming that balls with the same size are configured as an face-centered cubic (FCC) structure, upon exerting a pressure N, the sound speed is $\sim k^{1/2}$, and $k \sim N^{1/3}$; therefore, the sound speed is $\sim k^{1/2} \sim N^{1/6}$. The experimental results show that when the pressure is greater, the index is 1/6; when the pressure is lower, the index is 1/4, which implies $k \sim N^{1/2}$, and then $\sim k^{1/2} \sim N^{1/4}$, illustrating the limitations of the Hertz contact theory (see Figure 6.14).

In Figure 6.14, the diameter of the steel balls used in the experiments is 1/3 in in diameter. The slope of the solid line is 1/6, which is the calculation based on the Hertz theory; the slope of the dotted line is 1/4, which is the result of the conical contact model. The conical contact model is different from the Hertzian contact model because it assumes a zero radius of curvature at the point. The scatter points are the experimental results.

The contact stress can be deduced from the force chain deformation. For a granular system with constant volume that is subjected to shearing, the particles are passively adjusting their positions and the force chains subsequently rotate until unstable and

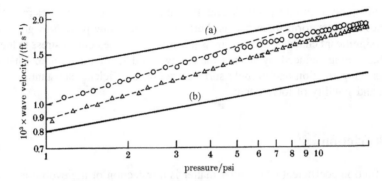

Figure 6.14: The sound speed as a function of the hydrostatic confining pressure in an FCC packing of 1/3-in diameter steel balls with low tolerance (O, ± 50 × 10⁻⁶ in) and high tolerance (△, ± 10 × 10⁻⁶ in). The solid lines have a slope of 1/6, indicative of a Hertzian contact, while the dashed lines have a slope of 1/4, indicative of a conical contact.

eventually bulk together. When the force chains rotate, dilatancy may occur. Due to the constraints of constant volume, force chain rotations are obstructed and the force chains become deformed, resulting in an elastic response. Supposing Δ is the total deformation of force chains, the contact deformation between particles of each pair is:

$$\alpha = \frac{\Delta}{N} = \frac{\Delta d}{L} \tag{6.8}$$

where N is the number of particles on a force chain and $N = \frac{L}{d}$, where L is the length of force chain and d is the particle.

The contact force between particles can be calculated by Hooke's law:

$$F = k\alpha = \frac{k\Delta d}{L} \tag{6.9}$$

The contact stress is:

$$\tau = \frac{F}{d^2} = \frac{k}{d} \tag{6.10}$$

The dimensionless contact stress $\tau d/k$ is:

$$\frac{\tau d}{k} = \frac{F}{kd} = \frac{\alpha}{d} \tag{6.11}$$

Therefore, $\tau d/k$ could be interpreted as the ratio of the particle contact deformation δ to the particle diameter d.

The particle surface friction is not included in the above analysis, but it affects the formation and stability of force chains. If the particle surfaces are smooth, force chains cannot bear any shear, i.e., a small amount of shear stress leads to breakage of the force chains. Therefore, the surface friction is an essential parameter in determining the mechanical properties of granular systems. For example, when the packing fraction is constant at 0.6, the shear rate γ is high, and the friction coefficient decreases from 0.5 to 0.1, the particles may not form force chains. Furthermore, the flow pattern of granular matter undergoes significant changes, and the stress transmits from the contact stress to the Bagnold stress, while being reduced by approximately two orders of magnitude. Therefore, the granular packing fraction and particle surface friction coefficients substantially affect the formation and stability of force chains.

6.4 Bulk friction

The bulk friction coefficient of granular matter is a reflection of the evolution of the force chain structure. For a three-dimensional plane shear flow shown in Figure 6.15, the thickness of the granular assembly is h and the stress P acts on the top plane with an area of S. The speed of the top plane is u, the friction force that the top plane experiences is F, and the bulk friction coefficient μ is:

$$\mu = \frac{F}{PS} \tag{6.12}$$

The shear rate dependence of μ can be empirically described by the Hershel–Bulkley model:

$$\mu = \mu_0 + sI^\alpha \tag{6.13}$$

The exponent α ranges from 0.3 to 0.4. I is the inertia number defined as:

$$I = \frac{\dot{\gamma}d}{\sqrt{P/\rho_p}} = \frac{ud}{h\sqrt{P/\rho_p}} \tag{6.14}$$

When the surface friction coefficient μ_s of the particle is variable, μ_0 also varies. For example, when $\mu_s = 0$, $\mu_0 = 0.06$; $\mu_s = 0.2$, $\mu_0 = 0.26$; $\mu_s = 0.6$, $\mu_0 = 0.4$. The power law of μ with I is held, as shown in Figure 6.16.

From Figure 6.16, it is found that μ is insensitive to shear rate in the lower regime of $I \leq 10^{-2}$. It indicates that the collision in the normal direction is irrelevant to the bulk friction coefficient in dense and slow granular flows. Note that μ is greater than μ_s; even when

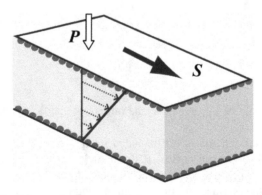

Figure 6.15: Schematic diagram of three-dimensional plane shear flow.

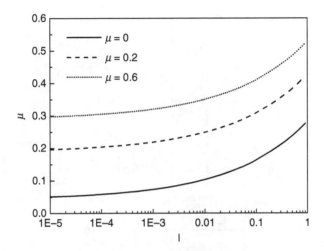

Figure 6.16: The bulk friction coefficient as a function of the inertial number in plane shear flows.

$\mu_s = 0$, μ is still greater than 0. The bulk friction coefficient μ then sharply increases around $I \leq 10^{-1}$.

6.5 Bulk restitution coefficient

Falcon et al. (1998) conducted an experimental study of the collision of a column of N beads ($N \leq 40$) with a fixed wall. The experimental device is shown in Figure 6.17. A column of N identical stainless steel beads, each 8 mm in diameter, is put inside a glass tube with an inner diameter of 8.1 mm. The number of beads may vary from $N = 1$ up to $N = 40$. Each bead has a tolerance of $\pm 4\,\mu m$ in diameter and $\pm 4\,\mu m$ in sphericity. Initially, the beads are at rest with no separation between them, and they are dropped from a height h above the wall. A sensor is connected to a numerical oscilloscope in order to record the collision with the column of beads. While the elastic wave is propagating through the force chains, the majority of the energy is dissipated on each of the contact surfaces. When the number of particles is sufficient, the energy of the particles is completely dissipated in less than 200 μs, so the overall restitution coefficient is almost zero, as shown in Figure 6.18.

For a collision between one bead and a plane at rest, the coefficient of restitution is usually dependent on the ratio of the bead's velocities after and before the collision. For N beads, the bulk restitution coefficient of the whole column is defined as:

$$\varepsilon_{\text{eff}} \equiv -\left(\sum_{i=1}^{N} v_i^f\right) / \left(\sum_{i=1}^{N} v_i^0\right) = -\frac{1}{N v_{\text{imp}}} \sum_{i=1}^{N} v_i^f \qquad (6.15)$$

Here, v_i^0, v_i^f denote the velocities of the ith bead after and before the collision, respectively. The velocities of all particles before the collisions are the same, namely, v_{imp}.

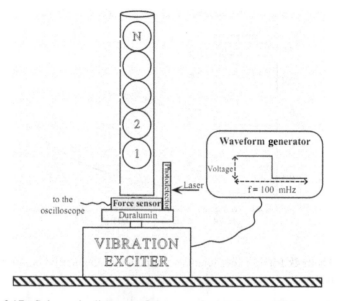

Figure 6.17: Schematic diagram of the experimental setup (Falcon et al. [8]).

6.6 Bulk elasticity

In addition to measurements of the interparticle force distributions, it is also important to measure the elastic responses of force networks as interferences are imposed on them. Currently, many studies have indicated that the unique mechanical laws of granular matter

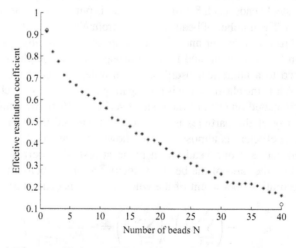

Figure 6.18: Restitution coefficient of the entire column as a function of N during the collision of N beads dropped from a height $h = 5.1\,\mathrm{mm}$. The restitution coefficient of a single particle is 0.92.

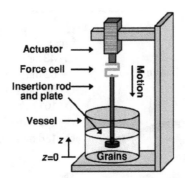

Figure 6.19: Granular penetrometer apparatus used for penetration measurements (Stone et al. [14]).

are related to the force chains. Stone et al. used the device shown in Figure 6.19. The packing fraction in the cylinder is about 0.59. The resistance force on the disc is measured as it is forced downwards in the granular matter. In order to reduce the effects arising from the connecting rod, its diameter (6.4 mm) is much smaller than the diameter of the disc.

The measurements are controlled and monitored by a computer using a commercial data acquisition board. The entire actuator and penetration assembly is rigidly attached to a wall. $z = 0$ represents the bottom of the vessel containing the grains. The actuator consists of a high-torque (0.44 Nm) computer-controlled stepper motor with a 200 mm range of linear motion and a distance-dependent voltage divider serving as a measure of the depth of penetration. The force cell has a capacity of 110 N acting in the directions of either tension or compression.

In Figure 6.20, typical measurements of F are shown for different filling levels, z_{max}, of the vessel. Each measurement consists of an initial linear regime where the penetration forces appear to be hydrostatic. For the deepest depth of grains, the hydrostatic regime is followed by a rollover to an almost depth-independent penetration force. This rollover is consistent with the stresses in the granular medium being described by a Janssen-like regime, where the sidewalls of the container support a significant portion of the weight of the grains. This feature is also in agreement with recent measurements examining this regime in other quasi-static systems. Near the bottom of the vessel, there is a rapid increase in the penetration force, which is due to the effect of the floor on the jamming of the grains in front of the penetrating plate. We note that the measurements for the largest values of z_{max}, namely, the deepest filling depth of grains, do not extend to the bottom of the vessel due to the limited range of motion of the actuator.

6.7 Correlation of force network with mechanical properties

In the abovementioned experiments, the bulk properties of granular systems are detected by exerting local perturbations. They can be qualitatively explained very well from the

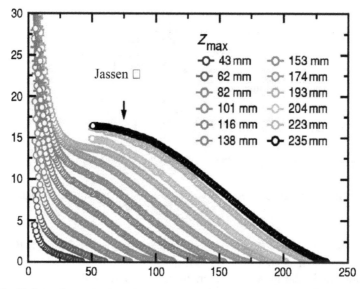

Figure 6.20: Height dependence of penetration force F for $d = 0.92 \pm 0.04\,\text{mm}$ and a $r = 12.7\,\text{mm}$ penetrating plate. The fill height of the vessel, z_{max}, increases from left to right for the data shown. The three different regimes are indicated.

perspective of the structure and its evolution of force networks. It indicates the importance of investigations on force chains, and further studies are currently being conducted in this area.

It is possible to make preliminary observations about the roles played by force chains under different strains. As illustrated in Figure 6.21, there is a range of strain over which the granular matter is elastic, i.e., the bulk deformation is recoverable. Inside the granular assembly, particles are displaced, relative to each other, so slightly that the displacement around the contact points is elastic. Accordingly, the spatial structure of force chains is intrinsically recoverable and leads to the granular material behaving like a solid. The underlying mechanism of granular plasticity is simple for low shear rates; it consists of topological events that rearrange particles, i.e., contacts are opened and closed by relative sliding or rolling. This leads to an irreversible evolution of force chains, namely, breakup and reformation of force chains. As the strain continues to increase, the yield stress is eventually attained, after which the irreversible plastic evolutions of force chains continue indefinitely under constant stress. Therefore, the granular matter flows like a fluid. When stress increases beyond the yield stress, the flow rate increases. Therefore, granular materials possess yield stress values, and belong to that category of complex materials termed 'Bingham fluids'.

How are force chains and mechanical properties of granular matter related? The contact network modification as a vital ingredient has been widely investigated as a potential link between the microscopic and macroscopic mechanics of most theories. The constitutive behavior of a granular assembly on a macroscopic scale has been derived from a microscopic scale description, taking a statistical description of the fabrics into account. Based on an analogy to fiber bundle models, a model has been proposed for the hardening of individual force lines during compression and their subsequent evolution.

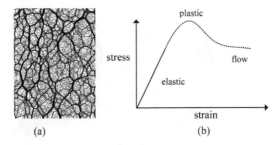

(a)   (b)

Figure 6.21: (a) Force chains and (b) macroscopic behaviors. Note that the schematic curve shown is for a monotonically increasing strain only. (a) A granular matter (here in two dimensional, in equilibrium) and internal force spatial distributed along chains. (b) The continuum rheological properties may be represented by the dependence of stress on strain and strain rate. In elastic regimes, recoverable force chain evolutions are dominant, and in plastic regimes, irreversible force chain evolutions are dominant.

The study of granular matter has been an active field over the past 20 years, and many intriguing phenomena are still hard to understand. Inspired by the scenario of the internal state of granular matter as revealed by our photoelastic experiments and numerical simulations, we postulated that dense granular matter is intrinsically multiscale: particle scale, force chains, and the granular assembly. The force chains play a vital role in determining the mechanical properties of granular matter. To date, there is no universal theory or model that can describe granular material behaviors for a wide range of conditions. Perhaps, the plausible theories can be established on the basis of force chains and their evolution dynamics.

6.8 Multiscale mechanics strategy

The size of a force chain is greater than the size of a single particle and smaller than the size of a granular system. It is not only related to the properties of granular materials (such as Young's modulus, Poisson ratio, etc.), but also impacted by the boundary and initial conditions. All of these determine the properties of a granular system, so the study of granular matter involves more than one physical scale structure and mechanism, belonging to multiscale mechanics: microscale particles, mesoscale force chains, and macroscale granular systems. Thus, multiscale spatiotemporal structures of force networks are the dominant feature for granular systems. Emphasis is being made on performing correlations between different scales, which is one of the focuses in the study of granular materials.

The correlation between force chains and their transformations with macroscopic properties has not yet been thoroughly studied. We introduced a mesoscale in granular matter, i.e., force chains bridging the two end scales. Force chains are determined not only by the material properties of the particles at microscale, such as the Young's modulus, Poisson ratio, and the static friction coefficient, but also by macroscopic parameters, such as the

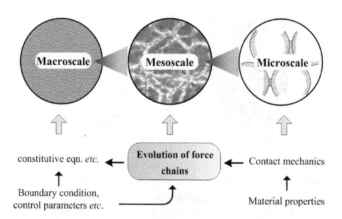

Figure 6.22: A multiscale method for dense granular materials. The role of force chains is specifically emphasized.

packing fraction, boundary conditions, and the loading history. Complicated transformations of force chains would dominate macroscopic mechanical properties of granular systems. A brief multiscale framework is illustrated in Figure 6.22. The obstacles, such as accurate description of force chains and interpretation of their statistical properties, remain. The related investigations represent a basic goal for granular physics in the near future.

Figure 6.22 briefly describes the relationship among these three. The microscale is mainly determined by the material properties of constituent particles; force chains are correlated with the contacting force, and evolve under the macroscale controls. The evolution law of force chains determines the macroscopic physical laws, such as the macroscale friction response and flexibility, the rheological relations, etc.

The roles of the three scales are different. Microscale research is the foundation; macroscale focuses on the applications, while mesoscale research is the bridge to link the basic research and application, and is also an important link to establish cross-scale correlation. However, it remains very difficult to establish the multiscale mechanics of granular materials. Currently, to the method for quantifying the force chains remains unclear, which, in turn, involves a series of basic questions: the force chains are stable in a certain time scale, for example, in the ring shear experiment; the duration of strong force chains is inversely proportional to the shearing rate, which is much greater than the contact collision time ($\sim 10^{-5}$ seconds) between microparticles; how then can the critical point of force chain duration be determined? How can the cut-off point to distinguish the strong and weak force chains be determined? How can the spatial distribution of the disordered force chains be described?

6.9 Characteristic time scales

Depending on their positions, particles in dense systems can be classified into two states, separated from each other or located within a contact connection. The two individual processes of a particle are resolved into: (1) separation or approach driven by the local pressure P, and after the duration of t_m, and (2) these particles contact, form a part of the network, and eventually, break after a lifetime of t_c. Therefore, the two processes in granular matter involve three time scales of very different natures: microscopic time t_m, lifetime of a contact t_c, and the time of the Rayleigh wave propagating along the particle surface, t_R.

The microscopic time t_m is:

$$t_m = \frac{d}{\sqrt{P/\rho}} \tag{6.16}$$

This represents the time it takes to displace a particle by one particle diameter d under the local pressure P, assuming the typical time scale of rearrangements. The variable ρ is the material density of the particles.

6.9.1 The macroscopic time scale t_c

For a granular system shearing at a rate of $\dot{\gamma}$, the correlated contact network rotates at a rate of $\sim \dot{\gamma}$, eventually becomes unstable, and self-destructs, at which point topological rearrangements of the particles occur. t_c would be represented as follows:

$$t_c = \frac{1}{\dot{\gamma}} \tag{6.17}$$

This relationship is linked to the lifetime of a contact. The quantity $\dot{\gamma}$ sets the fundamental time scale over which particle rearrangement events occur during the flow. For example, in the Bagnold stress formula, $\dot{\gamma}$ is the only relevant time scale in the problem that leads to a constitutive relation between the shear stress τ and strain rate $\dot{\gamma}$; $\tau = k\dot{\gamma}^2$, where k is independent of $\dot{\gamma}$.

For a dynamic contact, it is necessary to consider the elastic wave propagation across particles. It has been shown that the Rayleigh waves account for 67% of the radiated energy. It can be assumed that all of the energy is transferred by the Rayleigh waves with a speed of $v_R = (0.163\nu + 0.877)\sqrt{G/\rho}$, where G is the shear modulus.

The mesoscopic time scale t_R is as follows:

$$t_R = \frac{\pi R}{v_R} = \frac{\pi d}{0.163\nu + 0.877}\sqrt{\frac{\rho}{G}} \tag{6.18}$$

This is the time in which Rayleigh waves propagate along a particle surface, and thus, this represents the duration of elastic waves transmitting through a point of contact.

Campbell claimed that the timescales t_m and t_c relate to the stability of force chains.[15] However, in this work, we propose that these timescales may be related to the structural stability of the contact network, rather than the force chains, because force chains represent the pathways of force transmission. For example, in more slowly sheared granular materials, the contact network evolves slowly (i.e., the lifetime of the contact network $t_c = t_c = 1/\dot{\gamma}$ is much longer). However, the propagation paths of forces respond sensitively to any slight variations in contact force (in both magnitude and orientation), and transform instantaneously. This can be observed in many photoelastic experiments. The response time of a force chain t_{fc} is related to $N*t_R$, where N is the length of the force chain. For a typical length of $N = 10$, the time of $10t_R$ is still much shorter than t_m and t_c. Force chains instantaneously reflect the stress status in a granular assembly.

6.9.2 Three dimensionless numbers

The ratios of the three time scales can measure the relative importance of the three physical time scales. Three dimensionless numbers are proposed: inertial number I previously presented by Forterre and Pouliquen [9], Mach number Ma, and Deborah number De. The mechanical response of granular systems is related to the competition among these dimensionless numbers.

The inertial number I is the ratio between microscopic t_m and macroscopic t_c:

$$I = \frac{t_m}{t_c} = \frac{\dot{\gamma}d}{\sqrt{P/\rho}} \tag{6.19}$$

Small values of I correspond to a quasi-static granular bulk, in the sense that macroscopic deformation is slow compared to microscopic rearrangement. Large values of I correspond to rapid granular deformations. The dimensionless analysis tells us that one can either increase the shear rate $\dot{\gamma}$ or decrease the pressure P to switch from a quasi-static granular system to a rapid granular system in which the binary collisions are dominant. The variable I is also equivalent to the square root of the Savage or Coulomb numbers, which were introduced by some authors as the ratio of collisional stress to total stress.

The Deborah number De was first used in rheology. It is defined as the ratio of a relaxation time, characterizing the intrinsic fluidity of a material, to the characteristic time scale of an experiment. As indicated for granular systems, De can be obtained by the following equation:

$$De = \frac{t_R}{t_c} = \frac{\pi d \dot{\gamma}}{0.163\nu + 0.877}\sqrt{\frac{\rho}{G}} \tag{6.20}$$

When De is small, the contact network stays stable for longer, and the granular system stays quasi-static or deforms more slowly; As De becomes larger, contacts frequently open and close, and the system may yield toward flow. The meanings of De and I are similar, but the two ratios are from different characteristic time scales.

The Mach number Ma is the ratio between microscopic speed d/t_m and mesoscopic v_R:

$$Ma = \frac{d}{t_m v_R} = \frac{\rho}{(0.163\nu + 0.877)\sqrt{GP}} \tag{6.21}$$

This equation represents the competition between particle displacement speed and the speed of an elastic wave propagating through a contact connection. For rigid particles, G is infinitely large and $Ma \sim 0$. Any small amount of force imbalance in a contact transmits rapidly, and the contact network appears quasi-static.

Consider a model of sand grain system, where particle diameter $d = 1$ mm, Young's modulus $E = 5$ GPa, Poisson's ratio $\nu = 0.3$, density $\rho = 2650$ kg/m^3, and a load $P = 10^4$ Pa; as the system shears at a typical rate of $\dot{\gamma} = 20$ per second, the three characteristic times would be $t_m = 1.91 \times 10^{-4}$ second, $t_c = 5.00 \times 10^{-2}$ second, and $t_R = 3.98 \times 10^{-6}$ second. We further obtain $I = 3.82 \times 10^{-3}$, $De = 7.6 \times 10^{-5}$, and $Ma = 2.06 \times 10^{-4}$. For a typical

Table 6.2: Three rate processes and their characteristic time scales

Characteristic time			Correlation of characteristic time	
Life time of force chains	Construction and deconstruction of force chains	Elastic wave propagation time along particles	Meso/Macro	Formation/ propagation
$t_{fc} = \frac{1}{\gamma}$	$t_m = \frac{d}{\sqrt{P/\rho}}$	$t_R = \frac{d}{0.104\nu+0.558} \sqrt{\frac{P}{G}}$	$\frac{t_m}{t_{fc}} = \frac{d}{\gamma\sqrt{P/\rho}}$ $\frac{t_R}{t_{fc}} = \frac{d}{\gamma} \cdot \frac{1}{0.104\nu+0.558} \sqrt{\frac{P}{G}}$	$\frac{t_R}{t_m} = \frac{1}{0.104\nu+0.558} \sqrt{\frac{P}{G}}$

force chain length of $N = 10$, the force propagation time of $10t_R = 3.98 \times 10^{-5}$ second is much shorter than t_c and t_m. Therefore, the internal contact network is stable for a relatively long time period, and any imbalance at contacts due to perturbations of external forces would be rapidly balanced such that the sand system stays in the quasi-static regime.

These characteristic time scales combine with other field equations (such as the continuity equation, momentum, and energy equations), the statistical evolution equation of force chains, and the kinetics law of force chain formation and stress transmission, which finally may contribute to the construction of a multiscale framework. It is expected that the nonlinear dynamics properties and stress- and strain-localized problems in granular systems should be described.

References

[1] C. S. Campbell, 'Granular material flows: an overview', *Powder Technol.*, 162, 208–229 (2006).

[2] I. Albert, P. Tegzes, B. Kahng, R. Albert, J. G. Sample, M. Pfeifer, A.-L. Barabási, T. Vicsek and P. Schiffer, 'Jamming and fluctuations in granular drag', *Phys. Rev. Lett.*, 84, 5122–5125 (2000).

[3] J.-P. Bouchaud, M. E. Cates and P. Claudin, 'Stress distribution in granular media and nonlinear wave equation', *J. Phys. (France) I*, 5, 639–656 (1995).

[4] M. E. Cates, J. P. Wittmer, J.-P. Bouchaud and P. Claudin, 'Jamming, force chains, and fragile matter', *Phys. Rev. Lett.*, 81, 1841 (1998).

[5] C. S. Campbell, 'Granular shear flows at the elastic limit', *J. Fluid Mech.*, 465, 261–291 (2002).

[6] S. F. Edwards and R. B. S. Oakeshott, 'The transmission of stress in an aggregate', *Physica D*, 38, 88–92 (1989).

[7] J. M. Erikson, N. W. Mueggenburg, H. M. Jaeger and S. R. Nagel, 'Force distributions in three-dimensional compressible granular packs', *Phys. Rev. E*, 66, 040301 (2002).

[8] E. Falcon, C. Laroche, S. Fauve and C. Coste, 'Collision of a 1D column of beads with a wall', *Eur. Phys. J.B*, 5, 111–131 (1998).

[9] Y. Forterre and O. Pouliquen, 'Flows of dense granular media', *Annu. Rev. Fluid Mech.*, 40, 1–12 (2008).

[10] D. A. Head, A. J. Levine and F. C. MacKintosh, 'Distinct regimes of elastic response and deformation modes of cross-linked cytoskeletal and semiflexible polymer networks', *Phys. Rev. E*, 68, 061907 (2003).

[11] M. R. Horne, 'The behaviour of an assembly of rotund, rigid cohesionless particles—I, II', *Proc. R. Soc. London A*, 286, 62–97 (1965).

[12] G. Løvoll, K. J. Måløy and E. G. Flekkøy. 'Force measurements on static granular materials', *Phys. Rev. E*, 60, 5872–5878 (1999).

[13] T. S. Majmudar and R. P. Behringer, 'Contact force measurements and stress-induced anisotropy in granular materials', Nature, 435(1079), 1079–1082 (2005).

[14] M. B. Stone, R. Barry, D. P. Bernstein, M. D. Pelc, Y. K. Tsui and P. Schiffer, 'Local jamming via penetration of a granular medium', *Phys. Rev. E*, 70, 041301 (2004).

[15] M. Scheel, R. Seemann, M. Brinkmann, M. Di Michiel, A. Sheppard, B. Breidenbach and S. Herminghaus, 'Morphological clues to wet granular pile stability', *Nature Mater.*, 7(3), 189–193 (2008).

[16] J. Zhou, S. Long, Q. Wang and A. D. Dinsmore, 'Measurement of forces inside a three-dimensional pile of frictionless droplets', *Science*, 312(5780), 1631–1633 (2006).

[17] Q. Sun, F. Jin, J. Liu and G. Zhang, 'Understanding force chains in dense granular materials', *Int. J. Modern Phys. B*, 24(29), 5743–5759 (2010).

Chapter 7

Jamming and structure transformations

Granular materials are intrinsically athermal since their dynamics always occur far from equilibrium. For highly excited states, it can safely be assumed that only binary interactions occur. Constitutive relations can be determined by statistically tracking the repulsive force created in each interaction as the basis of kinetic theory. The majority of the theoretical considerations have been successfully applied to granular gases. However, for granular flows and granular solid states, they are still of technical relevance and innumerable continuum models have been presented, employing strikingly different expressions. Although the better ones achieve considerable realism when confined to the effects they were constructed for, these differential equations are more a rendition of complex empirical data, rather than a reflection of the underlying physics.

Widespread interest in granular media was aroused among physicists a decade ago, stimulated in large part by the intriguing fact that something as familiar as sand was still rather poorly understood. It brought forth a challenge to the classical thermodynamics and statistical physics. In the past decade, significant progress has been made on understanding the jamming phase diagram and the characteristics of the jammed phase. In this chapter, a brief overview on major advances is first presented. The pair-correlation function $g(r)$ is used to study the structural signature of the jamming transition of a granular assembly. Force networks may underlie the constitutive relations of both granular solids and granular flows and interstate transitions. However, it remains difficult to effectively describe the anisotropy of force networks. A two-time force-force correlation function for the normal force and positions is defined to characterize the evolutions of the system in the force space and position space. Meanwhile, a new pair-correlation function $g(r, \theta)$ is proposed to describe the characteristic lengths and orientations of force chains that are composed of particles with greater contact forces than threshold values. A formulation $g(r, \theta) \approx a(r) + b(r) \cos 2(\theta - \pi/2)$ is used to fit the $g(r, \theta)$ data. The characteristic lengths and orientations of force networks are then elucidated.

7.1 Introduction

As is known, there are a wide variety of disordered materials, including foams, gels, colloidal suspensions, and granular materials. All these systems exhibit a common nonequilibrium transition from a fluid-like phase to a solid-like phase, which is called the jamming transition (Figure 7.1). Three parameters control the characteristics of the jamming transition, namely:

$$\{T, \phi, \Sigma\} \tag{7.1}$$

where T is the temperature, ϕ is the volume fraction, and Σ is the shear stress.

Currently, physicists study the law of the jamming phase transition when one and/or two parameters change respectively. For example, the normal phase diagram of glass transition is in the $(1/\phi) - T$ plane, and it is divided into a jammed state region (such as glass) and an unjammed state region (such as liquid) by a transition line; however, the phase diagram of granular matter, foams, and gels lies in the $(1/\phi) - \Sigma$ plane, and the critical yield stress curve $\Sigma(\phi)$ divides the phase diagram into the jammed state region and the unjammed state region. For granular matter, the two states correspond to a granular solid and a granular liquid, respectively. In addition, there is another jamming phase transition in the $T - \Sigma$ plane, although there is no realistic system that corresponds to it so far.

Figure 7.1(a) shows the jamming phase diagram of a granular assembly composed of cohesionless particles (Liu and Nagel [1]). It can be seen that the characteristics of jamming phase transitions depend on the path taken to reach the phase boundary. For an isotropically jammed granular assembly, when ϕ is reduced to a certain point along the ϕ axis, bulk modulus and shear modulus approach zero. This transition point is called the J point, and the corresponding critical volume fraction is ϕ_c. The distance to the J point is defined as:

$$\Delta\phi = \phi - \phi_c \tag{7.2}$$

Theoretical and numerical studies have found that granular systems composed of frictionless particles exhibit some critical behaviors near J point, which, in some ways, resemble those seen with the second-order transition. For example, bulk modulus shows a power law scaling with $\Delta\phi$, and there are some divergent characteristic lengths when $\Delta\phi$

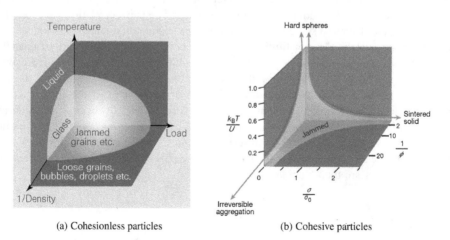

(a) Cohesionless particles   (b) Cohesive particles

Figure 7.1: Jamming phase diagrams of disordered materials.

Sources: Liu and Nagel [1] and Trappe et al. [2].

approaches zero. For the systems with finite size, the effect of the boundary on ϕ_c cannot be ignored, and ϕ_c is distributed within a certain range; while for the systems with infinite size, ϕ_c is a constant. In addition, the deformation of frictionless rigid particles is identically zero, and the granular systems always stay at the J point, that is, $\Delta\phi \equiv 0$.

Figure 7.1(b) shows the jamming phase diagram of a granular system composed of cohesive particles (Trappe et al. [2]). The three control parameters are:

$$\{\phi, k_B T/U, \sigma/\sigma_0\} \tag{7.3}$$

where U is the interaction energy between particles; σ is the shear stress; $\sigma_0 = k_B T/a^3$ is a function of temperature T and particle radius a, and is the characteristic stress exerted on the system. It can be seen that the granular liquid may be transformed into the granular solid by either increasing ϕ, U, or decreasing σ. The curvature of the phase boundary curves in Figure 7.1(b) differs significantly from those in Figure 7.1(a), and diverge at every corner. The maximum value of the $1/\phi$ axis corresponds to the irreversible aggregation, the maximum value of the $k_B T/U$ axis corresponds to the hard particle limit, while the maximum value of the σ/σ_0 axis (i.e., strong attraction among particles, high volume fraction ϕ) corresponds to a sintered solid. Furthermore, the characteristic yield stress diverges when $1/\phi \to 0$, while the system without cohesion between particles corresponds to a finite yield stress. Further studies need to be carried out for the comparison of the behavior of these two types of jammed systems.

The jamming phase diagram of disordered materials can provide a unified description of jamming phenomena, such as colloid-glass transition. For granular matter, the fundamental characteristic of jamming transition is the existence of relatively stable internal force networks, and that similar network structures are present in other disordered materials. It is only when cohesionless particles contact with each other that there is a mutual repulsion between them, and thus it is difficult to observe force chain networks. Whereas, for the system composed of cohesive particles, the network structure is stable and can be observed clearly, and this provides an opportunity to probe these force chains directly. Currently, study of the characteristics of the J point is still one of the difficult problems encountered in soft condensed matter physics.

7.2 Frictionless soft sphere systems

From the theoretical analysis point of view, systems of frictionless soft spheres are ideal to study the jamming transition. First, the J point of such a system is well defined. The jammed system has a finite shear modulus and yield stress when the boundary pressure $P > 0$; the shear modulus disappears when $P \to 0$, and thus unjamming transition occurs. At this stage, the system corresponds to zero pressure, zero shear, and zero temperature in the jamming phase diagram, that is, J point. Second, the system is marginally stable at J point, so the coordination number Z is close to the static value (i.e., $Z = Z_{iso}$). Third, under finite pressure, the mechanical and geometric properties of the jammed system present a peculiar power law scaling with $\Delta\phi$. Here, we introduce the peculiar geometrical and mechanical properties of frictionless soft sphere systems in the vicinity of J point.

7.2.1 Coordination number of an isostatic system

Coordination number Z is one of the key parameters to describe the structure of granular packing. A system is mechanically stable only when Z is greater than or equal to a critical value. The simplest jammed system is an isostatic system, in which the boundary pressure is isotropic and the system is stable, and the contact forces between particles can be determined by the condition of force balance and torque balance. Compared to other jammed granular systems, the value of Z in an isostatic system is relatively small.

Z is dependent on the spatial dimension d of a granular system and the roughness of particle surface. For a stable granular system composed of Nd-dimensional frictionless soft particles (e.g., N particles in a d-dimensional system), the total number of contact forces (the total degrees of freedom) is $NZ/2$ and the number of force balance equations (the total number of constraints) is Nd. Obviously, the contact force is solvable when $Z \geq 2d$, and $Z_{iso} = 2d$ corresponds to the isostatic value.

At J point, the pressure of the system is zero, there is no particle deformation, and the jammed conditions require that the distance between any two particles is exactly equal to the sum of their radii (just touching); therefore, the system has Nd position degrees of freedom and $NZ/2$ position constraints. Obviously, the system has a unique solution when $Z \leq 2d$, and $Z_c = 2d$ corresponds to the critical coordination number of J point. From the above analysis, we can see that the frictionless soft sphere system is isostatic at J point, that is, $Z_c = Z_{iso} = 2d$. Durian [3] carried out experiments of shearing in 2D liquid foam, in which the air bubbles can be regarded as soft particles, and the foam formed by the bubbles is a generalized 2D granular system. The results at J point indicated that:

$$Z_c = 4 \tag{7.4}$$

Brujic et al. [4] measured the coordination number of 3D particles in an emulsion by using a laser scanning confocal microscope technique; the results indicated the following:

$$Z_c = 6 \tag{7.5}$$

Durian also found that there was a scaling relation in 2D systems as follows:

$$(Z - 4) \sim (\phi - \phi_c)^{1/2}, \quad \text{i.e.} \ (Z - Z_c) = \Delta Z \sim \Delta\phi^{1/2} \tag{7.6}$$

Interestingly, subsequent studies have shown that this scaling relation is independent of the interaction potential and spatial dimension of particles.

In short, Z has a well-defined value at J point and shows a peculiar power law in the vicinity of J point. Taking into account that the mechanical properties of granular systems are sensitive to Z, it can be expected that the elastic modulus also show a similar scaling behavior in the vicinity of J point.

7.2.2 Elastics modulus

The system composed of frictionless particles is a typical marginally connected solid. Its bulk modulus B and shear modulus G depend on the type of interaction potential U between particles. In general, there are three kinds of interaction potentials, as follows.

(1) Hertzian interaction potential:

$$U \sim \delta^{5/2}, \, k \sim P^{1/3} \tag{7.7}$$

where k is the particle stiffness coefficient, δ is the overlap between particles, and P is the boundary pressure.

(2) Harmonic interaction potential:

$$U \sim \delta^2, \, k = \text{const.} \tag{7.8}$$

(3) General power-law interaction potential:

$$U \sim \delta^\alpha, \, k \sim \delta^{\alpha-2} \sim (\Delta\phi)^{\alpha-2} \tag{7.9}$$

Figure 7.2(a) shows the values of B and G for 2D granular systems with Hertzian interaction potential between particles (Durian [3]). Clearly, B satisfies $P^{1/3}$ scaling. If the effective medium theory can be extended to the granular system, its shear modulus will be scaled as $G \sim (\Delta\phi)^{\alpha-2}$. But, in fact, the results of numerical simulations show that the scaling relation of granular systems is:

$$G \sim (\Delta\phi)^{\alpha-3/2}, \, G \sim P^{2/3} \tag{7.10}$$

These relations apply toward Hertzian and Harmonic interaction potentials, respectively. It can be seen from Figure 7.2(b) that

$$G/B \sim \Delta Z^{1/3} \tag{7.11}$$

Here, $\Delta Z \sim \Delta\phi^{1/2} \sim P^{1/3}$.

7.2.3 Vibrational density of states

Density of states is an important concept in condensed matter physics. The vibrational spectrum of density of states of granular systems reflects information about the collective movement of particles; therefore, the vibrational mode and the related density of states are

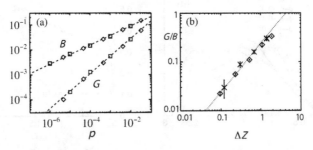

Figure 7.2: B and G with the boundary pressure P of a 2D granular systems with Hertzian interaction potential between particles (a), and the effect of ΔZ on B/G (b).

Sources: Durian [3] and O'Hern et al. [5].

important means to study the abnormal behavior of the system in the vicinity of J point. For a crystal, the low-frequency vibrational density of states satisfies the Debye relation:

$$D(\omega) \sim \omega^{d-1} \tag{7.12}$$

where d is the space dimension (i.e., $d = 3$ for three-dimension), and ω is the circular frequency of vibration.

If $D(\omega)$ deviates from the relation ω^{d-1}, it is said that the system exhibits anomalous behavior. O'Hern et al. [5] obtained the density of states spectrum of a 3D granular system with harmonic interactions by computer simulation, as shown in Figure 7.3. Apart from the J point (i.e., $\phi - \phi_c$ is large), $D(\omega)$ tends toward zero at low frequency, which is mainly consistent with the Debye scenario: when ϕ reduces to ϕ_c, $D(\omega)$ increases dramatically in the low frequency region, and a plateau appears. $D(\omega) \sim \omega^{d-1}$ is still valid at the left of the plateau, which is shown more clearly in Figure 7.3(b). Generally, taking the frequency at the left side of the plateau as a characteristic frequency, further studies have shown that the closer the system approaches the J point, the smaller ω^* is, and as $\omega^* \sim \Delta Z$ and $\Delta Z = 0$, $\omega^* = 0$ at J point. From the behavior of vibrational density of states $D(\omega)$, it can be seen that the closer the system approaches the J point, the greater the difference between the granular solid and an ordinary solid is.

We can wish to give a simple explanation for the plateau at low frequency of density of states. For an isostatic system, such as a jammed square (2D) or cube (3D) with side length l, there are a number of surface bonds of the order l^{d-1}. Since the system is isostatic, cutting the bonds at the surface can create, of order l^{d-1}, floppy zero-energy modes within the square or cube. The underlying floppy mode can make the system deform on the scale of l. Its frequency is $\omega_l = O(1/l)$. Thus, the $N_l \approx l^{d-1}$ mode can be produced in a box with $V_l \sim l^d$, and if the highest frequency of the mode is $\omega_l \sim l^{-1}$, we have:

$$\int_0^{\omega_l} D(\omega)d\omega \approx \frac{N_l}{V_l} \sim \frac{1}{l} \tag{7.13}$$

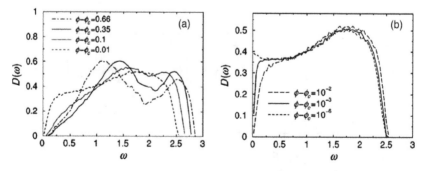

Figure 7.3: The density of states of 3D granular systems with Harmonic interaction potentials between particles.

Source: O'Hern et al. [5].

Assuming that $D(\omega)$ scales as ω^{α} at the low frequency, then:

$$(\omega_l)^{\alpha+1} \sim \frac{1}{l^{\alpha+1}} \sim \frac{1}{l} \tag{7.14}$$

Therefore $\alpha = 0$, that is, $D(\omega)$ is flat at low frequencies.

7.2.4 Microscopic criterion for stability under compression

The boundary pressure can have an impact on vibrational density of states. Assuming the contact length between two particles is s, the corresponding potential energy can be characterized by the normal stiffness and tangential stiffness coefficients. If the normal relative displacement is δ and the normal stiffness coefficient is k, then the normal force is $k\delta$. If the tangential relative displacement is x, the contact length increases by an amount proportional to x^2/s, and the work produced by the contact force is $\sim -x^2k\delta/s$, being equivalent to the tangential stiffness $\sim k\delta/s$. When a plane wave propagates in disordered solids, the transverse and longitudinal components of the contacts are of the same order of magnitude. Further the ratio of the two stiffness coefficient is $\sim \delta/s$, which is one of the measures of strain in solids. Therefore, the impact of compression on the plane wave under small strain can be ignored. It should be taken into account that there is no relative longitudinal displacement but only transverse displacement in the abnormal pattern, which is produced by the soft mode. Once the soft modes are deformed to generate abnormal patterns, they gain a longitudinal component of the order of ΔZ, while the transverse component remains on the order of unity. Therefore, the relative correction in the energy of the abnormal pattern is on the order of $-\delta/(s\Delta Z^2)$.

The anomalous modes become unstable, and the system yields when the relative correction reaches an order of unity. When approaching the J point, and $\delta/s \propto \Delta\phi$, the stability conditions follows the condition $\Delta Z > \Delta\phi^{1/2}$. This inequality must be satisfied for all the subsystems with sizes $L > l^*$, and this inequality extends the Maxwell criterion to the case of finite compression. For systems with smaller size, a violation of this inequality of fluctuation of the coordination numbers is allowed, for example, the fluctuation of the coordination numbers can be scaled as $\Delta Z \sim (\Delta\phi)^{1/2}$ in the vicinity of the J point.

7.2.5 Pair-correlation function

The pair-correlation function $g(r)$ of the frictionless soft sphere system exhibits many peculiar properties at the J point. The function $g(r)$ describes how the particle number density varies as a function of the distance r from one particular particle. For a 2D granular system, $g(r)$ is defined as:

$$g(r) = \frac{dN}{2\pi r dr}\frac{1}{\rho_0} \tag{7.15}$$

where dN is the particle number in the ring of $r \sim r + dr$, ρ_0 is the mean number density. For a 2D system, N particles distribute in the area S; $\rho_0 = N/S$.

It has been shown that a δ-function peak appears at the instant $r = a$ of $g(r)$, in the monodispersed system composed of particles each with a diameter of a. On the high side of this δ-function, $g(r)$ has a power-law decay:

$$g(r) \propto (r - a)^{-0.5} \tag{7.16}$$

This can be explained as the vestige of the marginal stability of the configurations visited before reaching ϕ_c. At J point, the second peak of the pair-correlation function splits into two peaks at $r = \sqrt{3}a$ and $r = 2a$, respectively.

Abate and Durian [6] experimentally studied the pair-correlation function near the zero-temperature jamming transition by using the upflows of gas through a mesh to control the movement of ball bearings (Keys et al. [7]). In the above systems, the kinetic energy of the ball bearings increases monotonically with the decrease of ϕ, and the kinetic energy becomes zero when $\phi \to \phi_c$. It was found that the height of the first peak of the pair-correlation function g_1 increases when approaching ϕ_c, and that there is a local maximum of g_1 at $\phi \approx 0.74$; while the coordination number and geometrical characteristics of the Voronoi cell do not change after $\phi > 0.74$. It has been found that the second peak of the pair-correlation function in the case of nonzero kinetic energy may correspond to a thermal vestige of the height divergence of the first peak at the zero-temperature transition. Zhang et al. [8] measured the pair-correlation function of a 2D bidispersed system at nonzero temperatures, shown in Figure 7.4. There is a maximum in g_1, and it can be explained as a thermal vestige of jamming transitions at the zero temperature. The numerical simulation shows that g_1 of the soft repulsive sphere system is diverse during the jamming transition at zero temperature:

$$g_1 \sim 1/(\phi - \phi_c) \tag{7.17}$$

The system softens to a finite maximum at nonzero temperatures. In addition, the maximum height of g_1 decreases with the increase in temperature, showing that the maximum value attained by g_1 is a structural feature of jamming transition. It is also apparent that this is a function of increasing density at fixed temperature, but not a function of temperature at fixed density or pressure.

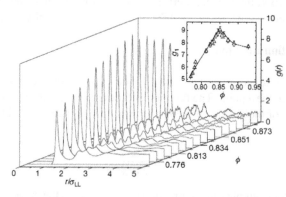

Figure 7.4: The pair-correlation function of a 2D system composed of bidispersed colloidal particles at different values of ϕ.

Source: Zhang et al. [8].

7.3 Frictional soft sphere systems

The system of frictionless soft spheres has a well-defined jamming point. The system is isostatic at the jamming point (i.e., $Z_c = Z_{iso}$), and the change of the coordination number near the J point demonstrates characteristics that are hybrids of the first-/second-order characteristics: below the J point, $Z = 0$; at the J point, Z jumps from 0 to $2d$; above the J point, $(Z - Z_c)$; and the mechanical quantities of the jammed system exhibit a power law scaling with $(\phi - \phi_c)$, which is independent of dimensions, interaction potential, and polydispersity. In contrast with the frictionless soft sphere system, ϕ_c and Z_c of frictional soft sphere system are not unique and both depend on the system's friction coefficient and the preparation history. The granular system of frictional soft spheres is hyperstatic (i.e., $Z_c > Z_{iso}$) at the J point. Its mechanical properties scale with $(Z - Z_{iso}^\mu)$, while $(Z - Z_c)$ scales with $(\phi - \phi_c)$. The hyperstatics of frictional soft sphere systems at the J point imply that the contact forces between particles cannot be uniquely determined by the geometry of the system.

7.3.1 Critical coordination number

We focus on the coordination number of frictional soft sphere systems at zero pressure. On the one hand, the condition that particles are barely in contact is the same as that of frictionless soft sphere systems, that is, $ZN/2$ restriction is applied on N_d particle position coordinates, and requires the condition $Z \leq 2d$. On the other hand, there are $ZN_d/2$ components of contact forces and N_d force balance equations and torque balance constraint equations in a frictional system, which requires $Z \geq d + 1$. Combining these two constraints, the range of the coordination number of the frictional soft sphere system is:

$$d + 1 \leq Z_c < 2d \tag{7.18}$$

The range of the coordination number based on the above constraints' statistics includes all the force configurations satisfying the balance conditions of force and torque, but these configurations do not always satisfy the Coulomb criterion $f_t/f_n \leq \mu$ where f_n and f_t are the normal force and tangential force components at the contact, respectively, and μ is the static friction coefficient. In the limit of $\mu \to \infty$, the Coulomb criterion is satisfied naturally, but when the coordination number is at the J point Z is close to the lower limit:

$$Z_{iso}^\mu = d + 1 \tag{7.19}$$

where Z_{iso}^μ is the isostatic value of a frictional soft sphere system. For the case of finite values of μ, the Coulomb criterion must be considered, in which the critical coordination number at the J point is the function of μ, that is, $Z_c(\mu)$. Indeed, numerical simulations and experiments have found that when μ increases gradually from 0, Z_c does not jump from $2d$ to $d + 1$, but rather to Z_c. However, Z_c is not a clearly defined curve. It depends on the preparation history of a granular system. Numerical simulations have found that the static systems with different Z and ϕ at J point can be generated by changing the quenching rate. Therefore, it is not only the impact of the Coulomb criterion, but also the impact of the preparation history that should be considered when analyzing the coordination number of frictional systems at J points.

7.3.2 Generalized isostaticity

Numerical simulations have found that some of the contacts lie at the Coulomb yield threshold in the slow prepared systems, that is, f_n and f_t are no longer independent and they satisfy $f_t = \mu f_n$. The contacts where the normal contact force and tangential contact forces satisfy the above relation are the so-called fully mobilized contacts (FMC), and the number of FMC will affect the microscopic criterion of stability. Assuming that the average number of FMC per particle is n_m, then there are $ZNd/2$ contact force components, while there are $n_m N$ FMC constraints in addition to Nd force balance equations and $d(d-1)N/2$ torque balance equations constraints. Following these discussions on the microscopic criterion for stability of smooth particles systems, the generalized microscopic criterion of stability can be introduced for frictional system as:

$$Z \geq (d+1) + 2n_m/d = Z_{iso}^m \qquad (7.20)$$

Thus, the system with $n_m = d/2(Z - Z_{iso}^\mu)$ is an isostatic or marginal system, while the system with $n_m < d/2(Z - Z_{iso}^\mu)$ is a hyperstatic system. Usually, the system with the maximum number of FMC is called a generalized isostatic system, and it can be implemented by slow preparation. Numerical simulations have found that the values of n_m and Z in the generalized isostatic line for $P \to 0$ satisfy the above constraints. Figure 7.5 shows the slow prepared systems with different values of μ. It can be seen that in the generalized isostatic line,

$$\mu \to \infty, \ Z \approx d+1, \ n_m = 0 \qquad (7.21)$$

$$\mu \to 0, \ Z \approx 2d, \ n_m \approx d(d-1)/2 \qquad (7.22)$$

In generalized isostatic systems, a large number of FMC lead the system to obtain almost zero energy deformations, resulting in a large number of low-energy excitations. Therefore, the density of states in the low-frequency region increases greatly, and this has been verified with measurements of the vibrational density spectrum.

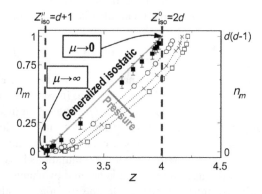

Figure 7.5: The $n_m - Z$ phase diagram under different boundary pressures.

Source: van Hecke [9].

7.3.3 $Z - \phi$ phase diagram

The volume fraction ϕ at the J point is not unique for frictional systems, but depends on the friction coefficient and preparation history. Song et al. [10] established the statistical volume description of jammed states of frictional systems by proposing a partition function of volume ensembles, and obtained the $Z - \phi$ phase diagram describing the volume fraction of the system with different friction coefficients, as shown in Figure 7.6. In terms of volume ensembles, random loose packing (RLP) and random close packing (RCP) correspond to the limits of compactivity $X = \infty$ and $X = 0$, respectively, and the density of ground state of jammed granular matter is:

$$\phi_{RCP} \approx 0.634, \quad \phi_{RLP}(Z) \approx \frac{Z}{(Z + 2\sqrt{3})} \tag{7.23}$$

where Z is a function of the friction coefficient. Figure 7.6 shows the statistical interpretation of the RLP and RCP limits:

(1) The RCP limit provides the maximum ϕ of disorder packings. It uses the friction coefficient to characterize the ground state of hard sphere systems. When μ change from 0 to ∞, the RCP state changes correspondingly, namely the RCP limit in the phase diagram is not a single point.
(2) The RLP limit gives the minimum value of ϕ for a given Z, along the RLP lines; the value of ϕ of RLP decreases as μ increases from 0 to ∞.
(3) Systems that are intermediate between the RLP and RCP limits have a finite X (i.e., the yellow area shown in Figure 7.6).

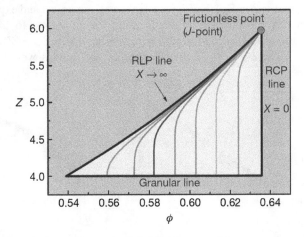

Figure 7.6: The $Z - \phi$ phase diagram of rigid sphere packing.

Source: Song et al. [10].

7.3.4 Characteristic frequency of density of state and the modulus ratio G/K

A large number of experiments and numerical simulations have shown that Z deviates from Z_c in frictional soft sphere systems. $Z - Z_c$ exhibits square root scaling law with $(\phi - \phi_c)$ as:

$$Z - Z_c = Z_0(\phi - \phi_c)^{1/2} \tag{7.24}$$

where Z_0 is a fitting factor. This is equivalent to $(Z - Z_0) \sim P^{1/3}$ for Hertzian contacts. It is noteworthy that $Z - Z_{iso}^{\mu}$ at a finite pressure does not scale with $(\phi - \phi_c)$ since Z_c of a frictional soft sphere system is different from the isostatic value $Z_{iso}^{\mu} = d + 1$.

The calculation for the 2D vibrational density spectrum of frictional systems shows that the characteristic frequency $\omega*$ can be scaled linearly with $Z - Z_{iso}^{\mu}$, as shown in Figure 7.7(a). Similarly, the study of shear modulus G and bulk modulus K of a 2D frictional system shows that G/K also scales with $Z - Z_{iso}^{\mu}$, as shown in Figure 7.7(b). These findings suggest that the scaling law of frictional soft sphere systems is determined by the distance from the point of isostaticity, rather than the distance to the jamming.

7.4 Jamming of other disordered systems

The earlier results aim at understanding static granular systems composed of soft spheres, combined with practical situations, such as irregular particle shapes, nonstatic shear, and so on. Some conclusions would be helpful to the study, to understand jamming in other disordered systems.

7.4.1 Jamming of nonspherical particles

Statistical arguments about the degrees of freedom imply that the coordination number of an ellipsoid at the J point is $Z = Z_{iso} = d(d + 1)$. However, numerical and experimental results show that Z of slightly nonspherical ellipsoids at the J point is close to $2d$, which suggests that a granular system of weak ellipsoidal particles has $(N/2)(Z_{iso} - Z)$ soft modes. For spherical particles, where $Z \rightarrow 2d$, the number of soft modes would be $d(d - 1)N/2$.

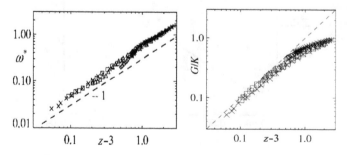

Figure 7.7: The scaling of $\omega*$ and G/K of a 2D frictional system with $Z - Z_{iso}^{\mu}$.

Source: Somfai et al. [11].

For frictionless soft sphere systems, $d(d-1)N/2$ soft modes can be obtained accurately, corresponding to free rotation spheres. The characteristics of soft modes in the nonspherical hypostatic systems under finite pressure can be studied by observing the vibrational density spectrum.

Jamming transition of 2D and 3D ellipsoid systems was carefully studied in our experiment. Since each ellipsoid can be characterized by five degrees of freedom, the isostatic coordination number of an ellipsoid is $Z_{iso}^{ellipsoid} = 2 \times 5 = 10$. Our numerical simulations found that the average coordination number changes from $Z_{iso} = 10$ to $Z_{iso} = 6$, successively, when the ellipticity $\varepsilon = c/a$ (c and a are the semiminor axis and the semimajor axis of an ellipse, respectively) changes from a very small value (i.e., the shape is a distant sphere) to $\varepsilon = 1$. Calculations show that when $Z < Z_{iso}$, there are $N_c(Z_{iso} - Z)/2$ directions in the phase space of these particles. The particles' coordinates can change along these directions without changing the interaction energy among these particles, which means that each static state configuration has many zero-frequency modes. The numerical calculation of the density of states of an ellipsoid particle system found that the number of zero modes is consistent with the theoretical expectation, that is, the number of zero mode decreases with the increases of $\delta_\varepsilon = \varepsilon - 1$. Thus, with the increase in Z, an increasing number of nonzero modes would appear. Zeravcic et al. (2009) found that there were two obvious separate subbands in the density of states for small values of ε (such as $|\delta_\varepsilon| < 0.17$), as shown in Figure 7.8(a). The modes in the upper band of the density of states are very similar to the modes found in the spherical particle systems, while the initial frequency ω^* is a characteristic frequency, which can be scaled as follows:

$$\Delta Z = Z - 6 \sim |\delta_\varepsilon|^{1/2} \tag{7.25}$$

Furthermore, when ε is small, the low-frequency band mainly comprises the rotation modes, while its upper frequency scales with $|\delta_\varepsilon|$ linearly. When $\varepsilon > 0.17$, these two bands merge into a mixed band, as shown in Figure 7.8(b).

It is worth mentioning that despite the nonspherical system being hypostatic, the jamming of the frictionless ellipsoid systems is very similar to the jamming of the frictional soft spherical systems. Near the jamming point, there are large numbers of soft modes which do

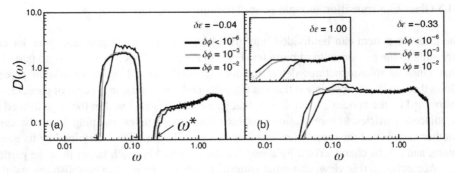

Figure 7.8: The density of states spectrum of a 3D system composed of ellipsoid particles.

Source: Zeravcic et al. [12].

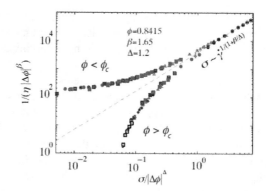

Figure 7.9: The rheological behavior of a foam system under shear.

Source: Olsson and Teitel [13].

not affect the stiffness of the system either for the frictional soft spherical system or the weak ellipsoid system.

7.4.2 Jamming of foams under shear

Bubbles in liquid foams can essentially be treated as frictional soft spheres with viscosity due to surface tension. The jamming phase diagram of the foam system can be explored by studying the rheological properties under constant shear stress and constant shear rate. Olsson and Teitel [13] modeled the shear rheological data near the frictional J point collapse, as shown in Figure 7.9, where $\eta = \sigma/\dot{\gamma}$ is the effective viscosity, σ is the shear stress, and $\dot{\gamma}$ is the shear rate. σ can be scaled as:

$$\sigma = \dot{\gamma}^{0.4} \tag{7.26}$$

This applies toward large values of $\dot{\gamma}$ near ϕ_c. This is a simple paradigm of deriving macro-scale rheological properties by using a microscopic model (Olsson and Teitel [13]).

7.4.3 Glass-like transition of rigid granular fluid

The granular system can be divided into granular solid, granular liquid, and granular gas, according to the motion state of particles. For a granular liquid composed of rigid particles, what kinds of microstructures determine its viscosity? What kind of relaxation process allows them to flow? One view is that the viscosity and the relaxation process of particles are determined by the system's local nature, that is, a particle is enclosed in the cage formed by its adjacent particles; the relaxation process relates to particles escaping from the cage. Another view is that the mechanical stability of a granular liquid is determined by its global nature, and can be characterized by a length scale l^* which is much larger than the particle size. According to this view, abnormal patterns theory can be used to establish the stability criteria of a granular liquid. Figure 7.10 shows that the stability of a granular liquid can be described by the $\delta Z - h$ phase diagram (Brito and Wyart [14]), where h is the average gap

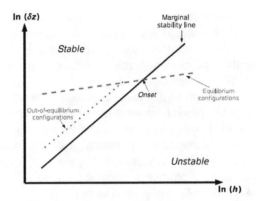

Figure 7.10: The $\delta Z - h$ phase diagram of a granular liquid composed of rigid particles.
Source: Brito and Wyart [14].

between adjacent particles. In Figure 7.10, the critical stability line divides the stable configurations from the nonstable configurations and satisfies the following relation:

$$\Delta Z > P^{-1/2} \sim h^{1/2} \tag{7.27}$$

The dashed line (Figure 7.10) corresponds to the equilibrium configurations with different ϕ (or h). The smaller ϕ is, the greater h is, and the more unstable configurations exist. The variable h decreases gradually with the increase in ϕ, and the stable configuration line begins crossing the critical line of stability. The value of ϕ at the crossing point is ϕ_{onset}. The viscosity increases rapidly at large values of ϕ, and the configurations become more stable. The system is out of equilibrium eventually for a finite quenching rate, depicted as the dotted line in Figure 7.10. The position deviating from the equilibrium line depends on the quenching rate, and the nonequilibrium line is close to the borderline of stability at a fast quenching rate limit. It is worth mentioning that the critical stability is the basic characteristics of glass, which include the peculiar behavior, the abnormal patterns near-zero frequency, and the short-time dynamics. For example, the mean square displacement of particles can scaled as:

$$< \delta R^2 > \sim (\Delta \phi)^{3/2} \tag{7.28}$$

This equation can be contrasted with the parameter $\Delta \phi^2$ that is measured in crystals. This prediction equation (7.28) has been confirmed (Somfai et al. [11]).

7.5 Structural transformation in a frictional system

Jamming is the physical process by which some materials, such as granular materials, glasses, foams, and other complex fluids, become rigid with increasing density. The phase diagram for granular systems depends on temperature, load, and density, as illustrated in Figure 7.1. Jamming can occur only when the density is high enough. In particular, a key question concerns how stability can occur when the packing fraction ϕ increases from below to above a critical value ϕ_c, for which there are just enough contacts to satisfy the conditions of mechanical stability. For frictionless soft spheres, there is a well-defined jamming transition indicated by the J point on $1/\phi$ axis, which exhibits similarities to a (unusual) critical phase transition. We can then unjam the system either by raising temperature T or by applying a shear stress Σ. The phase diagram raises some interesting questions: a jammed granular material has a lower yield stress when random motions (i.e., thermal fluctuations) are present. In this work, the jamming/unjamming of a frictional sphere system is simulated with discrete element simulation. Both mechanical and structural criteria are used to determine the J point, including the pair-correlation function $g(r)$, force–force correlations function C^n, C^t, and a position–position correlation function C^{xy}. The scaling law of the boundary pressure P with $(\phi - \phi_c)$ was studied as well.

7.5.1 Numerical simulations

Five thousand round disks in 2D are generated in a square cell of 4×4 m^2. To avoid crystal packings, a bidispersed distribution is used. The ratio of radii between small and large particles is 1:4, and the number ratio is 1:1. The constituent particle density is 2,600 kg/m^3, the normal and tangential stiffness are both 1.0×10^8 N/m, and the friction is $\mu = 1 \times 10^{-4}$. The contact potential among particles is harmonic and the acceleration due to gravity is ignored.

To study the jamming behavior along the $1/\phi$ axis, at the initial state, the radii of 5,000 particles is small so that they loosely distribute in the cell. Obviously, the assembly stays in the unjammed state. The particle radius is proportionally increased to ensure the value of ϕ is enlarged at a small step of 1×10^{-4}, while keeping the radius ratio constant, until the system enters the jammed state. At each time step of small increments of ϕ, particles reach new positions after sufficient time.

Figure 7.11 shows the variation of boundary pressure P with the packing fraction ϕ. As $\phi < 0.835$, P is always close to zero; at $\phi = 0.8353$, P sharply increases to 2017 Pa and then continuously increases. It should be noted that as ϕ increases by a step of 1×10^{-4} at 0.8356, the value of P suddenly drops, which indicates the phenomenon of internal stress relaxation, as shown at Point A. This phenomenon is called stick and slip, which is common in granular materials.

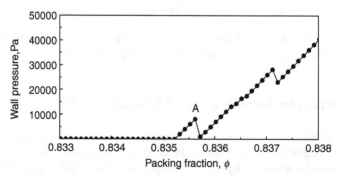

Figure 7.11: Boundary pressure vs. packing fraction in the isotropically compressed system.

7.5.2 Pair-correlation function $g(r)$

For a granular system, the relative positions of constituent particles obey some statistical rules, for example, the probability of finding the center of a particle of a given distance from the center of another particle. For short distances, this is related to how the particles are packed together. For example, consider hard spheres. The spheres cannot overlap, so the closest distance between two centers is equal to the diameter of the spheres. However, several spheres can be touching one sphere, after which a few more can form a layer around them, and so on. Further away, these layers become more diffuse, and so for large distances the probability of finding two spheres with a given separation is essentially constant. In that case, it is related to the density: a more dense system has more spheres, thus it is more likely to find two of them within a given distance. The pair-correlation function $g(r)$ accounts for these factors by normalizing against the density; thus at large values of r, it attains 1, indicating uniform probability.

In this work, d_r is 0.0002, around 1/100 of the mean diameter. The calculated $g(r)$ distribution is shown in red in Figure 7.12. The gray curve is the correspond $g(r)$ for a Gaussian size distribution of a granular assembly. It can be seen that $g(r)$ has a series of peaks for the Gaussian distribution, the first peak called g_1 exists at distance $r = <d>$, where $<d>$ is the mean diameter. However, for the bidispersed system, besides g_1 being consistent with the peak obtained in the Gaussian distribution, there exist another two peaks around the first peak in the Gaussian distribution. This indicates two small particle contacts and two large particle contacts, respectively. It can further be seen that $g(r)$ has an oscillating shape characteristic of any disordered medium.

Figure 7.13 shows that as ϕ increases from the unjammed phase, the height of the first peak of $g(r)$, g_1, increases. It indicates that more particles come in to contact. But when

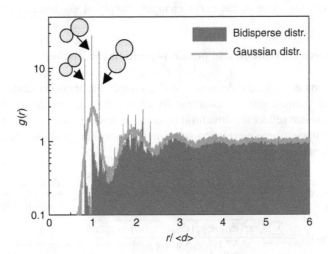

Figure 7.12: Pair-correlation function $g(r)$ with the distance r. The radius r is given in the unit of the average dry particle diameter $<d>$. The height of the first peak of $g(r)$, g_1, is compared with that of the Gaussian distribution.

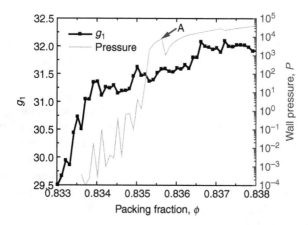

Figure 7.13: Comparison of the structural signature g_1 with the force measured along the boundary pressure P. Point A is indicated in Figure 7.11.

$\phi > 0.8342$, g_1 increases slowly and starts exerting forces on boundaries. Strictly speaking, the system enters the jammed state, where $\phi_c = 0.8342$. Note that $P(\phi)$ fluctuates largely around ϕ_c. From these observations, $g_1(\phi)$ is sensitive to granular jamming. One interesting observation is that when $\phi > 0.8376$, at some value of ϕ, g_1 appears as a local maximum. This implies that the internal structure may change at this value of ϕ, but P is still very large. The system may transform from one jammed state to another jammed state, for example, along with increasing values of ϕ, local jamming occurs. To illustrate the detail of the onset of the jamming, the pressure is also plotted on the logarithmic scale. Below the jamming point, the force fluctuates around the noise level, around a few Pa. As the value of ϕ increases further, the slope of the curve changes sharply at ϕ_c, indicated by the arrow in Figure 7.13.

7.5.3 Force–force correlation and position–position correlation

For high fractions $\phi > \phi_c$, the decreases in $P(\phi)$ imply the transformation of one jammed state to another jammed state. In isotropically compressed assemblies studied, such drops in boundary pressure reflect the structural changes in force networks. To characterize such evolutions, the force–force correlation function and position–position correlation function are introduced, and are defined as:

$$C^n(\phi_0, \phi) = \frac{\sum_{ij} |f^n_{ij}(\phi_0)||f^n_{ij}(\phi)|}{\sum_{ij} |f^n_{ij}(\phi_0)|^2}, \quad C^t(\phi_0, \phi) = \frac{\sum_{ij} |f^t_{ij}(\phi_0)||f^t_{ij}(\phi)|}{\sum_{ij} |f^t_{ij}(\phi_0)|^2},$$

$$C^{xy}(\phi_0, \phi) = \frac{\sum_i \sqrt{x_i^2(\phi_0) + y_i^2(\phi_0)}\sqrt{x_i^2(\phi) + y_i^2(\phi)}}{\sum_i |x_i^2(\phi_0) + y_i^2(\phi_0)|}$$

(7.29)

where, taking $\phi_0 = 0.8420$ as the reference state, the sum examined over all pairs of particles (i, j) corresponds to a spatial average. f^n_{ij} and f^t_{ij} are the normal and the tangential forces;

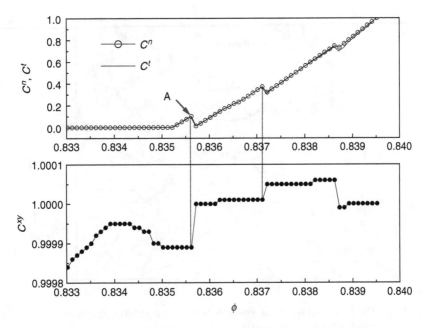

Figure 7.14: Variation of C^n, C^t, and C^{xy} along with increases in ϕ.

x_i and y_i are the center positions of particle i. As shown in Figure 7.14, after $\phi = 0.835$, the force correlation function C^n and C^t exhibit small jumps, corresponding to microslips, revealing the unusual occurrence of bursts in the reorganization of the force network. During these bursts, the energy due to the tangential interaction decreases, whereas the energy due to the normal interaction increases. But, $C^{xy}(\phi_0, \phi)$ is nearly unchanged, which indicates that the geometric configuration remains unchanged. During the flat regime of $C^{xy}(\phi_0, \phi)$, $C^n(\phi_0, \phi)$ and $C^t(\phi_0, \phi)$ increase linearly with increases in the value of ϕ, which indicates that at each jammed state, the magnitudes of normal and tangential forces increase approximately linearly. At some critical value of ϕ, $P(\phi)$ rapidly increases and $C^n(\phi_0, \phi)$, $C^t(\phi_0, \phi)$, and $C^{xy}(\phi_0, \phi)$ also change simultaneously. Although $C^n(\phi_0, \phi)$, $C^t(\phi_0, \phi)$, and $C^{xy}(\phi_0, \phi)$ change, $C^{xy}(\phi_0, \phi)$ changes with a magnitude of $\sim 5 \times 10^{-5}$. This is because the particle position changes very slightly. Much larger changes are observed in $C^n(\phi_0, \phi)$ and $C^t(\phi_0, \phi)$, on the order of $\sim 10^{-1}$. The force chains at the state denoted as Point A, and after an increment of 1×10^{-4}, are shown in Figure 7.15. It can be seen that although the particle positions are nearly constant, the force networks change considerably.

7.5.4 Unjamming process

The scaling of the shear modulus G and bulk modulus K plays a central role in connecting the disordered nature of the response to the anomalous elastic properties of systems near jamming. To understand why this disorder plays such a crucial role globally, mechanical

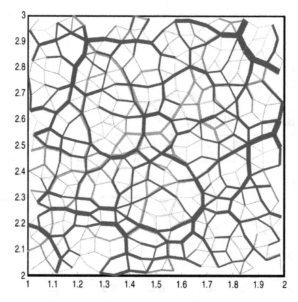

Figure 7.15: The force networks at Point A and after a small increment (1×10^{-4}) of ϕ, inredandblue, respectively.

response of a collection of particles that act through short-range interactions consider the local motion of a packing of spherical, soft frictionless spheres under global forcing.

The unjamming process is realized by relaxing the obtained jammed states. The packing fraction becomes smaller very slowly, that is, the step of ϕ is -1×10^{-5}. It can be seen that for each jammed state, the boundary pressure P is almost linearly reduced during the unjamming process; once it has reached zero, the value of ϕ is ϕ_c. Note that for different jammed states, the ϕ_c is different, but is distributed within a narrow range from 0.83791 to 0.83820, as shown in Figure 7.16. It may be caused by the scale size effect, that is, when the number of particles is high, the range of variation in ϕ_c is narrow.

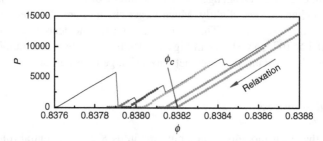

Figure 7.16: Boundary pressure P with ϕ during unjamming process. For each jammed state, the value of P is almost linearly reduced during the relaxations.

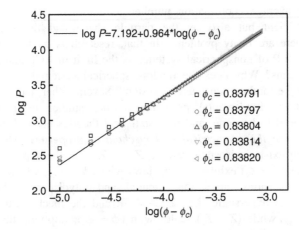

Figure 7.17: Scaling laws of boundary pressure P with $(\phi - \phi_c)$. The exponent index is 0.964. Symbols denote different jammed states.

Earlier data have shown that the bulk and shear modulii of frictionless granular systems obey scaling laws with the distance to jamming point $\Delta\phi \equiv \phi - \phi_c$, $P \sim (\phi - \phi_c)$. For frictional granular systems in this work, a similar scaling law is obtained. Figure 7.17 shows that when $\log P \sim \log(\phi - \phi_c)$ the exponent index is 0.964. Currently, there is no satisfactory physical understanding about the exponent due to the influence of friction.

7.6 Conclusions

The states of disordered materials can transit from the unjammed state to the jammed state, and such jamming transition is realized by changing the thermodynamic quantities (i.e., temperature or density) and loads (i.e., the stress applied to the sample). Granular materials with repulsive contact interactions at zero temperature and zero shear stress provide a good sample for understanding the nature of jamming transitions and jammed states. Large numbers of experimental and numerical studies have shown that granular matters, while appearing to be simple (e.g., particles could be frictionless, spherical, and deformable), exhibit many complex geometrical and mechanical behaviors near the J point. So far, extensive studies have been conducted on the jamming transitions of granular systems, but no satisfactory understanding of the physical nature of jamming transitions has been obtained. The following problems remain open.

(1) *Concept of random loose packing (RLP) and random close packing (RCP):* Song et al. [10] used the terms of volume ensemble to provide a statistical interpretation about the RCP and RLP limits of frictional soft sphere systems, namely the RCP and RLP limits corresponding to the ground state density of jammed granular matter at the limit of $X = 0$ and $X = \infty$, respectively. There exists some controversy about the control and realization of X; further studies need to be conducted. Further, the volume

fraction and average coordination number of a system not only depends on the friction coefficient but also on the particle shape and the preparation history. Currently, there are many problems; in that, research is unclear about the concept of RCP and RLP of nonspherical systems. Is the limit of RCP and RLP the same in spherical systems? Why does the random spherical system have a smaller volume fraction than the random nonspherical system? Several other questions exist. In order to answer these fundamental problems, further studies using experiments and simulations need to be done, and new statistical theories need to be developed.

(2) *Relation between the geometrical and mechanical properties:* The frictionless soft sphere system exists in isostaticity (i.e., $Z_c = Z_{iso}$) at the J point. The mechanical quantities and $(Z - Z_c)$ exhibit scaling laws with $(\phi - \phi_c)$, which is independent of the system dimensions, interaction potential, and polydispersity. The frictional soft sphere system is hyperstatic (i.e., $Z_c > Z_{iso}$), and the mechanical properties scales with $(Z - Z_{iso}^{\mu})$, while $(Z - Z_c)$ scales with $(\phi - \phi_c)$, implying that the mechanical properties are not uniquely determined by the geometry of the system. The study of the density of state of 3D spherical systems shows that the subband of weakly deformed spheroids corresponds to translational modes, and its characteristic frequency ω^* can be scaled with $(Z - Z_{iso}^{sphere})$ rather than $(Z - Z_{iso}^{ellips})$, after which the change of shape from spherical to a weakly deformed ellipsoid can be seen as a smooth perturbation. Under a large boundary pressure, the translational band and the rotational band would merge into one mixed band, and its characteristic frequency ω^+ scales with $(Z - Z_{iso}^{ellips})$, while there are $(Z - Z_{iso}^{ellips}/2)$ soft modes in the system with $Z < Z_{iso}^{ellips}$. At the spherical-shaped limit, these modes correspond to local rotations, and these modes are delocalized when present away from the spherical shape. In short, these systems appear to be simple, but exhibit new physics. The close link between the geometrical and mechanical responses needs further study.

(3) *Nonlinear behaviors and fluctuations near jamming:* So far, most studies have only focused on the average quantities and linear responses of disordered materials. The mechanical responses are primarily affine distant from the J point, while the mechanical responses of disordered systems become nonaffine near the point J. In addition, some quantities, such as coordination number, modulus, and others, of different configurations in the finite-sized systems are different from one another. There are still some questions which we cannot answer clearly, such as how does one understand these fluctuations near jamming? What is the nature of the nonlinear yielding of the system near jamming? In particular, similar problems in the system with finite shear and finite temperature warrant further study.

(4) *General picture of jamming:* Jamming provides a framework for understanding the mechanics of disordered systems. The study on stable frictionless soft spheres shows that these systems exhibit complex spatial organization and unusual mechanical properties at the limit of isostatic jamming. An important task for the next few years is to expand the framework of jamming and extrapolate it to the more general systems, while taking into account the impact of shear and temperature. However, how to extend the picture of a frictionless soft sphere to the systems with general interactions and nonspherical shapes is still an open question. The physics appearing in more general scenarios of jamming is worthy of further experimental and theoretical explorations.

References

[1] A. J. Liu and S. R. Nagel, 'Jamming is not just cool any more', *Nature*, 396, 21–22 (1998).

[2] V. Trappe, V. Prasad, L. Cipelletti, P. N. Segre and D. A. Weitz, 'Jamming phase diagram for attractive particles', *Nature*, 411, 772–775 (2001).

[3] D. J. Durian, 'Foam mechanics at the bubble scale', *Phys. Rev. Lett.*, 75, 4780–4783 (1995).

[4] J. Brujic, C. Song, P. Wang, C. Briscoe, G. Marty and H. A. Makse, 'Impact of a projectile on a granular medium described by a collision model', *Phys. Rev. Lett.*, 98, 248001 (2007).

[5] C. S. O'Hern, L. E. Silbert, A. J. Liu and S. R. Nagel, 'Jamming at zero temperature and zero applied stress: the epitome of disorder', *Phys. Rev. E*, 68, 011306 (2003).

[6] A. R. Abate and D. J. Durian, 'Approach to jamming in an air-fluidized granular bed', *Phys. Rev. E*, 74, 031308 (2006).

[7] A. S. Keys, A. R. Abate, S. C. Glotzer and D. J. Durian, 'Measurement of growing dynamical length scales and prediction of the jamming transition in a granular material', *Nature Phys.*, 3, 260–264 (2007).

[8] Z. Zhang, N. Xu, D. T. N. Chen, P. Yunker, A. M. Alsayed, K. B. Aptowicz, P. Habdas, A. J. Liu, S. R. Nagel and A. G. Yodh, 'Thermal vestige of the zero-temperature jamming transition', *Nature*, 459, 230–233 (2009).

[9] M van Hecke, 'Jamming of soft particles: geometry, mechanics, scaling and isostaticity', *J. Phys.: Condens. Matter*, 22, 033101 (2010).

[10] C. Song, P. Wang and H. A. Makse, 'A phase diagram for jammed matter', *Nature*, 453, 629–632 (2008).

[11] E. Somfai, M. van Hecke, W. G. Ellenbroek, K. Shundyak and W. van Saarloos, 'Critical and noncritical jamming of frictional grains', *Phys. Rev. E*, 75, 020301(R) (2007).

[12] Z. Zeravcic, N. Xu, A. J. Liu, S. R. Nagel and W. Van Saarloos, 'Excitations of ellipsoid packings near jamming', *Europhys. Lett.*, 87, 26001 (2009).

[13] P. Olsson and S. Teitel, 'Critical scaling of shear viscosity at the jamming transition', *Phys. Rev. Lett.*, 99, 178001 (2007).

[14] C. Brito and M. Wyart, 'Geometric interpretation of previtrification in hard sphere liquids', *J Chem. Phys.*, 131(2), 024504 (2009).

Chapter 8

Point loading response and shear band evolution

In dense granular systems, external loadings selectively transmit along the pathways connected by interparticle contacts, and accordingly, an internal force-chain network is formed. Its structural evolution leads to the mechanical nonlinearity and irreversible deformation of each of such systems, as well as the transition between solid-like and liquid-like phases. In this chapter, by using the discrete-element method (DEM), the propagation of a point loading onto the surface of a static granular assembly is studied. It is a simple and convenient way to analyze the structure of force networks and the induced stress anisotropy. For a uniaxially compressed granular system, three conditions to define a force chain are proposed, and the chain length distribution is found in the form of a power law. The shear band is a well-known failure mode of granular materials and occurs when a granular material system has been subjected to large shearing deformation. The development of shear band is studied, particularly by monitoring internal parameters, such as coordination number, particle rotation angle, and force-chain types. The energy transformation is analyzed as well.

8.1 Point loading transmission

The response to a localized force provides a sensitive test for models of stress transmission in granular solids. Photoelastic experiments have shown that when a disordered granular system is subjected to a localized loading, the stress distribution exhibits a near-parabolic envelope, which is wider than what is predicted by continuum elasticity. Goldenberg and Goldhirsch [1] have conducted extensive research in this field. They reported that elasto-plastic models have been challenged by theories and experiments that suggest a wave-like (hyperbolic) propagation of the stress, as opposed to the elliptic equations of static elasticity. Their results indicated that in large systems, the response was close to that predicted by isotropic elasticity, whereas for small systems, it is strongly anisotropic. The particle friction and polydispersity affect the range of elastic response. The larger the surface friction, the more extended the range of forces for which the response is elastic, and the smaller the anisotropy. An increase in the degree of polydispersity decreases the range of the elastic response.

Point loading

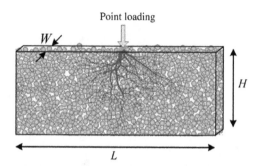

Figure 8.1: Schematic diagram of a point loading onto a static granular assembly.

8.1.1 Numerical experiment setup

A total of 10,000 particles of each of three sizes are used: $d = 0.01$, 0.008, and 0.006 m, respectively. They fall freely under gravity and are packed in a Hele–Shaw cell with dimensions of $L = 2$ m, $H = 1$ m, and $W = 0.01$ m, as shown in Figure 8.1. For constituent particles, density $\rho = 2000$ kg/m^3, friction coefficient $\mu = 0.2$, and the normal and tangential stiffness coefficients are both 1×10^8 N/m. Some surface particles are removed to obtain a relatively flat surface. A point loading P is exerted on one of the surface particles after the system arrives at a static state.

The point loading is $P = 100 m_{max} g = 5.2 \times 10^{-2}$ N, which is 100 times the maximum particle weight. P must be very small, so as not to cause the plastic deformation of the granular surface. An example of the interparticle force distribution obtained in a run is presented in Figure 8.2. It can be seen that some particles bear a greater force and form quasi-linear strong force chains. Due to the cumulative particle weight, strong force chains increase with the increase in depth. A number of strong force chains are cross-linked to form a few significant arches, so that they could effectively carry external loadings, while the remainder of the particles stay in the pores of the arch and bear very weak forces. In order to calculate the response, the forces entirely due to gravity (without the applied force) were vectorially subtracted at each contact, as shown in Figure 8.2(b). The force transmission pathways can then be clearly observed. Note that the thickness of a force chain is proportional to the

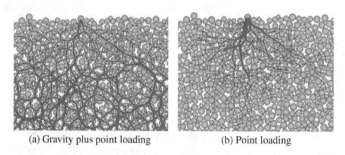

(a) Gravity plus point loading   (b) Point loading

Figure 8.2: (a) Force networks under gravity plus a point loading and (b) solely at a point loading. $\mu = 0.2$, $P = 5.2 \times 10^{-2}$ N.

magnitude of the contact force. In order to highlight the force chains, the scale in the two figures are different.

8.1.2 Point loading transmission

The friction coefficient μ would not only directly influence the sliding and rotation of particles, but would also significantly influence the structure and evolution of force chains, and eventually, could affect the response of the granular system. In this chapter, $\mu = 0.2$ is constant, but we selected seven different loading positions on the surface to observe the differences among force chains, as shown in Figure 8.3.

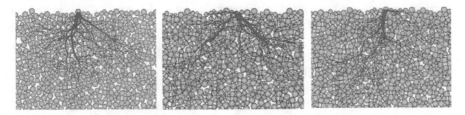

Figure 8.3: Three force networks generated by three different loading positions.

It can be seen from Figure 8.3 that the oriented force chains result in non-uniform stress components in a localized area. This kind of anisotropy is called induced anisotropy, which is a function of external loading and not the intrinsic nature of the material. It is found that the transmission direction of interparticle forces is dependent to the adjacent particles. In other words, the force transmission direction observes the particle centroid as the center and radiates outward through contact, and not necessarily perpendicularly downward. Therefore, the calculation of stress should take the loading particle on the surface as the center to establish a polar coordinate, and then the corresponding average radial stress σ_{ij} in a local area is expressed as:

$$\sigma_{ij} = -\left(\frac{\phi}{\sum_{N_p} V^{(p)}}\right) \sum_{N_p} \sum_{N_c} \left|x_i^{(c)} - x_i^{(p)}\right| n_i^{(c,p)} F_j^{(c)} \tag{8.1}$$

where ϕ is the packing fraction in the local area, $V^{(p)}$ is the volume of particle p, $x_i^{(c)}$ is the coordinate of a contact point, $x_i^{(p)}$ is the coordinate of particle centroid, $n_i^{(c,p)}$ is the unit normal vector from the centroid to the contact point, and $F_j^{(c)}$ is the contact force. The results are shown in Figure 8.4. Note that the distance is normalized with the maximum particle diameter.

It can be seen that the force transmission angles (the angle between the force direction and the horizontal direction) is different in the three loadings. The stress in one direction is often greater than in another direction (see Figure 8.4(a)), but in many runs, the probability to be on the left or right is roughly equal (see Figure 8.4(c)).

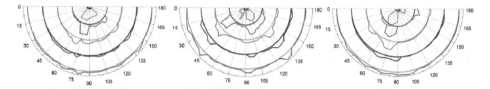

Figure 8.4: Stress distributions in the three loading positions, as shown in Figure 8.3.

Figure 8.5 shows the standard deviation δ of the stress at different distances r from the loading point. The value of δ decreases along with r, and this follows an exponential law. It implies that the elastic–plastic theory is not suitable for describing the mesoscale mechanical properties characterized by the particle size, and can only describe the macroscopic mechanical properties of a granular system whose dimensions are greater than the size of a single particle.

The localized loading P onto a static granular system is matches the description of a concentrated loading per unit length, and the stress can be predicted by the Flamant solution:

$$\sigma_R = -\frac{2P}{r}\cos\varphi \tag{8.2}$$

where r is the radius and φ is the angle between the loading and the vertical direction.

In Figure 8.6, the horizontal axis represents the angle between the measurement point and the vertical direction, while the vertical axis represents the stress of the measurement point. The dashed line denotes the Flamant solution, while the open squares denote the value at $r = 6$ obtained by this work. It can be observed that the fluctuations are greater, mainly due to the limited number of particles. Basically, these data are consistent with the theoretical values.

From Figures 8.5 and 8.6, it can be seen that there exist obvious scale effects in granular systems. It is consistent with the Saint-Venant's principle that is regarded as a statement on

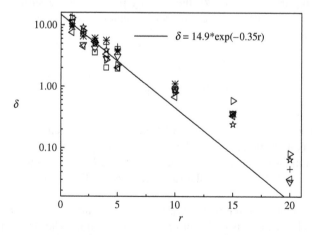

Figure 8.5: The standard deviation δ of stress distribution. The distance r is normalized with the maximum particle diameter.

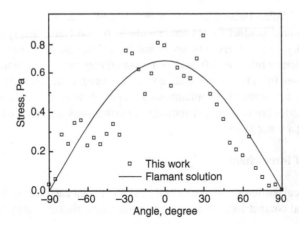

Figure 8.6: Comparison of stress distributions with the Flamant solution; $R = 5$ in this case. R is normalized with the maximum particle diameter.

the asymptotic behavior of the Green's function by a point-load: "*... the strains that can be produced in a body by the application, to a small part of its surface, of a system of forces statically equivalent to zero force and zero couple, are of negligible magnitude at distances which are large compared with the linear dimensions of the part*".

8.2 Force network under uniaxial compression

As illustrated in Figure 8.7, 12,400 particles in 2D are densely packed in a box. The particle diameters are 0.4, 0.5, and 0.6 mm, respectively, and the size distribution is uniform. The material properties of constituent particles are as follows: density $\rho = 2650\,\text{kg/m}^3$, elastic

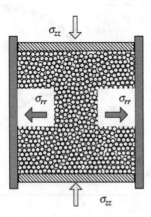

Figure 8.7: Schematics of a two-dimensional granular system under uniaxial compression. The axial stress σ_{zz} is imposed on the upper and lower boundaries; the left and right walls are fixed, and the induced radial stress is σ_{rr}.

modulus $E = 100\,\text{MPa}$, Poisson's ratio $v = 0.3$, and friction coefficient $\mu = 0.3$. The normal and tangential contact forces are calculated by the Hertz theory and the Mindlin–Deresiewicz theory, respectively. The upper and lower boundaries move inward to compress the granular system while the left and right boundaries remain stationary. The packing fraction ϕ increases from 0.827 to 0.886, and the induced axial and radial stresses, σ_{zz} and σ_{rr}, respectively, are increased simultaneously. When ϕ reaches a certain value, the upper and lower boundaries are maintained stationary. Further runs are required to ensure that the system arrives at a static state.

8.2.1 Criteria of force chains

When two spherical particles with the radii R_1 and R_2 contact each other, the contact radius is a, then the normal contact force F and the corresponding elastic energy W are:

$$F = \frac{4}{3} E^* \left(R^* \right)^{1/2} \delta^{3/2}; \quad W \approx \frac{1}{5R^*} \left(\frac{3R^*}{2E^*} (1 - v^2) \right)^{2/3} F^{5/3} \quad (a/R^* \to 0) \qquad (8.3)$$

where $1/R^* = 1/R_1 + 1/R_2$, and $1/E^* = (1 - v_1^2)/E_1 + (1 - v_2^2)E_2$, where E_1 and v_1, and E_2 and v_2 are the elastic moduli and the Poisson's ratios of particle 1 and particle 2, respectively, and δ is the overlap between particles.

Figure 8.8 shows the probability density distribution of F and W along with the dimensionless contact force $F/\langle F \rangle$. The distribution of contact force as depicted in the inset is well-known; for a static granular system, a peak appears in the vicinity of $\langle F \rangle$, and the force on only 40% of the contact points is greater than $\langle F \rangle$; more importantly, the distributions before and after $\langle F \rangle$ are very different; when $F < \langle F \rangle$, the probability density distribution increases according to a power law along with the increase in F; when $F > \langle F \rangle$, the probability density distribution decreases according to an exponential law along with the

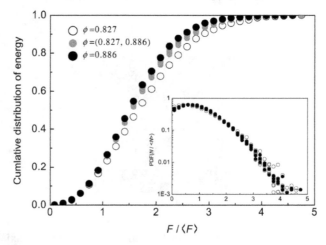

Figure 8.8: Cumulative distribution of elastic energy. The inset indicates the distribution of contact forces.

increase in F. It can be observed from the accumulative diagram of elastic energy that the 40% of the contact points account for 80% of the elastic energy.

The contact force distribution shows that the force chains with $F > \langle F \rangle$*** are responsible for most of the energy in the system and are certainly related to the macroscopic properties of granular systems. According to the magnitude of the contact force, chain forces can be divided into strong force chains and weak force chains. They both exist in the granular system, and transform into each other under certain conditions.

8.2.2 Lateral pressure coefficient

The lateral earth pressure is the pressure that soil exerts in the horizontal plane. To describe the lateral pressure, a lateral pressure coefficient K is used. K is the ratio of lateral (horizontal) pressure to vertical pressure ($K = \sigma_{rr}/\sigma_{zz}$). Thus, the horizontal earth pressure is assumed to be directly proportional to the vertical pressure at any given point in the soil profile. Janssen assumed K to be a constant because of the fact that the axial pressure tends toward a saturation value exponentially, but K may depend on the soil properties and the stress history of the soil.

In general, K is directly related to the internal friction angle of granular materials, which, in turn, is related to the interparticle friction and the geometrical embedding and interlocking among particles. Both of them are generalized as the internal friction: the rougher the soil particle surface, the greater the angles and subsequently, the greater the internal friction angle of granular materials. From the perspective of force chains, the corresponding force-chain structure becomes more stable at this time, so the internal friction angle of granular materials reflects the responses of the internal force-chain structure and the strength to the external loadings.

Figure 8.9 shows the variation of the calculated value of K along with ϕ. It can be seen, when $\phi < 0.848$, K is relatively large, and when $\phi > = 0.848$, $K = 0.54$. In many previous experiments, $K \approx 0.4$–0.5 in three-dimensional laterally confined soils, which is slightly smaller than the results obtained in this work. Comparing the internal force-chain network in Figure 8.14(b), when $\phi = 0.827$, K is most likely not dense enough. The force chains are not fully developed, and their directions are randomly distributed. When extruding the upper and lower boundaries, ϕ increases. The system continuously becomes dense, new force chains generate, and the directions of many force chains gradually shift to the direction of σ_{zz}; when $\phi > = 0.848$, K stably tends to 0.54, and the system becomes sufficiently dense. In other words, the force chains become fully developed and almost do not change.

By analogy with the definition of K, a new lateral energy ratio $S = W_{zz}/W_{rr}$ may be defined, where W_{zz} and W_{rr} are the radial and axial elastic energies, respectively. According to equation (8.3), when the deformation is extremely small, $S \sim F^{5/3} \sim K^{5/3}$; taking into account the definition of S, the following can be obtained:

$$S = 1.33K^{1.7} - 0.29 \tag{8.4}$$

The variation of S along with ϕ is shown in Figure 8.13. By comparing the curves of $S - \phi$ and $K - \phi$, it indicates that the lateral energy ratio of strong force chains may be determined by the coordination number. K is an essential macromechanical parameter of granular systems, and S represents the energy distribution of strong force chains. The overlap

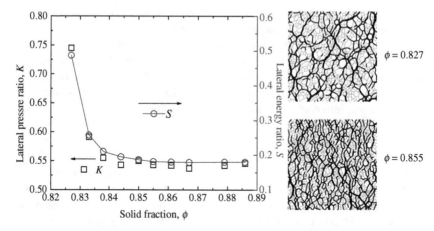

Figure 8.9: Lateral pressure coefficient compared to S. The right insets show the corresponding force chains.

of the two curves illustrates that the strong force chains play a decisive role in the macromechanical properties of granular systems.

8.3 Shear bands

Under asymmetric external loadings, the deformation of a granular system is usually highly concentrated in a narrow shear band, which leads to the failure of the material. This may further lead to the phase transition from solid-like to liquid-like behavior. The width of such a shear band is usually about six to eight times the particle diameter. Radiography has been used since at least the early 1960s to observe the changing positions of lead shot markers in, typically, plane strain soil models and elements. Lead shots absorb X-rays much more than the soil mass, leaving clear images on the X-ray film. Displacements of a somewhat regular grid of shot from one film to the next can be interpreted as indications of a continuum in the displacement field and a corresponding strain field.

Since the occurrence of a shear band involves quite a few intrinsic unknowns, such as, time of shear band formation, density distribution, and thickness and its evolution, a complicated instrumentation, e.g., X-ray computed tomography technique, is usually required to explore its characteristics. Nevertheless, the two essential variables in the identification of shear bands are the thickness and density inside the localized zone. In physical models of district normal faulting, shear-band-induced reduction in sand density due to dilatancy and corresponding reduction in X-ray absorption allows observation of fault development. For the evolution of void ratios within shear bands of triaxial sand samples using the X-ray computed tomography technique, the local density inside shear band approaches a limited value of the void ratio. The local void ratio was reported to be much lower than the global measured value.

Over the past 40 years, in-depth research on shear bands has been conducted from the aspects of theoretical analysis, experimental testing, and numerical simulation. In theoretical analysis, the initial constitutive models cannot effectively predict the occurrence of shear bands. Some new theories, such as the gradient plasticity theory, bifurcation theory, and others, have thus been proposed, with which a lot of engineering problems were solved; however, the understanding of shear bands is still very limited. A non-local damage theory was proposed to extend the traditional localized damage mechanics model to the non-localized damage mechanics model. In other words, the stress on a point is not only related to the strain changing history of this point, but also affected by the interaction of materials in a certain neighborhood around this point. The research on granular mechanics has immediately abandoned the assumptions of continuum, and treats granular matter as a collection of discrete particles, and places more attention on the mesoscale structure, such as force chains.

In the aspect of experimental tests, observations of shear bands have been conducted on Karlsruhe dry sand under the plane strain state; the shear band inclination, width, and the evolution laws with different particle diameters were carefully studied. Subsequently, three-dimensional photography was used to measure the strain field inside and outside the shear bands. With the application of photoelastic technology, interparticle forces and force chains can be observed by interference fringes. In the aspect of numerical simulations, Cundall conducted shear band simulations on soil particles by using the DEM. The results are similar to the ones obtained from the disc photoelastic experiments. A large number of numerical simulations have been carried out so far to measure volumetric strain, shear band thickness, and particle displacements' rotation.

During the formation and development of shear bands, a distinct solid-like to liquid-like transition feature is exhibited, so the shear bands can be regarded as processes of jamming and unjamming as defined in physics. Many advances in jamming studies may shed light onto the mechanics of shear bands. In this work, the discrete-element software PFC2D is used to simulate the formation and evolution of shear bands under biaxial compression. The change of the packing fraction, coordination number and force-chain networks during the shear band development process is analyzed.

8.3.1 Macroscopic phenomena

The granular assembly is 0.2 m high, 0.1 m wide, and consists of 4325 round particles in two dimension. The particle diameter is subject to a uniform probability distribution in the range of 0.0046 to 0.0075 m. The particle surface friction coefficient $\mu = 0.5$, the normal and tangential stiffness are both 5.0×108 N/m. The computing time step is $\Delta t = 10^{-6}$ second. The acceleration due to gravity is neglected. A confining pressure 2 MPa is maintained on the left and right boundaries. A constant vertical loading speed of 0.1 mm/s (i.e., axial direction) is imposed on the upper and lower boundaries. In order to keep the assembly in a quasi-static condition, the loading speed should be slow enough, which can be measured by comparing the boundary work and elastic energy of particle samples. For the elastic deformation stage, the two should be equal. Figure 8.10 shows the stress–strain curve during loadings. Three velocity fields are shown for the corresponding strain. The deviatoric stress is $\sigma = \sigma_1 - \sigma_2$, where σ_1 is the axial stress and σ_2 is the horizontal confining pressure. The axial strain is ε.

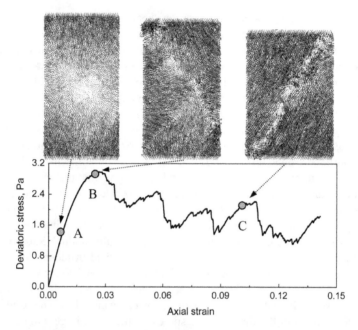

Figure 8.10: The stress–strain relation of the system. The insets depict the velocity fields.

It can be observed from the above figure that the assembly experiences a complete elastic–plastic process. When $\varepsilon < 2.0\%$ (point A), the granular system exhibits elasticity. Since the loading is imposed on the upper and lower boundaries at the same time, the velocities on the upper and lower boundaries are the maximum, and the velocity field shows a clear central symmetry. With the increase in ε, the assembly enters the plastic phase. Before $(\sigma_1 - \sigma_2)$ achieves the peak value, many localized microbands appear. When $\varepsilon = 2.8\%$ (point B), the granular system reaches its peak strength, and then enters the softening stage, i.e., with stress decreasing sharply. With the continuous increase of ε, the displacement of particles changes from central symmetry to longitudinal symmetry, and shows a clear X-shaped band. When $\varepsilon = 4.2\%$, the granular system enters the critical stress stage. The particle velocity has a clear direction, and develops gradually into a major shear direction, leading to the eventual formation of a large shear band. During this process, the deviatoric stress appears, first with a small value, and then with larger fluctuations. It can be seen from the velocity field of $\varepsilon = 10\%$ (point C) that particles within the shear band rotate greatly, and particles beyond the shear band move in bulk like rigid bodies. According to the distribution of the velocity field, the thickness of the shear band is roughly concentrated within 8 to 10 particle diameters.

For the orientation of the shear band, there are two traditional theoretical solutions, namely, the Mohr–Coulomb solution and the Roscoe solution. The Mohr–Coulomb solution is usually considered as the upper limit, while the Roscoe solution is the lower limit. Since neither of them is appropriate for some experiments, Arthur has proposed a definition that the angle is the average of the two solutions. In this work, the width of shear band in the lower

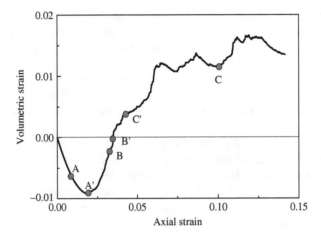

Figure 8.11: The volume strain variation. The corresponding points A, B, and C are shown in Figure 8.14.

left corner is narrower than the upper right corner, but the angle of the medial axis remains at about 55.5°. When $\mu = 0.5$, the corresponding internal friction angle is $\varphi = 26.6°$, and then the Mohr–Coulomb solution is 58.3°. For the Roscoe solution, if the corresponding dilation angle uses the slope value at the initial stage of axial strain and volumetric strain, the angle of the shear band is 53.8°; thus, it can be observed that the obtained angle in this work is more consistent with the Arthur solution.

Figure 8.11 shows the evolution of the volume strain ε_v ($= \varepsilon + \varepsilon_2$, where ε_2 is the horizontal strain). It can be observed that ε_v first reduces to a negative value, the packing fraction ϕ increases, and the system shows an elastic strain at this time. When $\varepsilon \approx 2\%$ (i.e., $\varepsilon_v = -0.8\%$), the system achieves the most dense state (point A'), and the packing fraction achieves its maximum value. After $\varepsilon > 2\%$, ε_v begins to increase, and the system enters the plastic stage. When $\varepsilon = 3.5\%$ ($\varepsilon_v = 0$), the volume strain returns to the initial state (point B'). After $\varepsilon > 3.5\%$, ϕ decreases gradually, ε_v appears to show a slight fluctuation, but the slope of change remains the same at this time. When $\varepsilon = 4.2\%$, the system enters a critical state, and εv increases continuously, but exhibits a larger fluctuation (point C'). It is observed that force chains within the assembly continuously collapse and reform.

8.3.2 Mesoscale analysis on shear bands

Over the past 20 years, granular matter as a new type of condensed matter has aroused the concern of physicists. An initial consensus has been reached on some fundamental aspects, such as the jamming phase diagram in granular matter. When transformed from fluid-like to solid-like phases, it is known as the jamming transition, namely, from an un-jammed state to a jammed state. The reverse process is called unjamming transition. The process is controlled with temperature T, packing fraction ϕ, and shear stress Σ. At present, scientists have studied the jamming transition when one (or two) parameters change, on the plane of $(1/\phi) - \Sigma$, and divided by the critical yield stress $\Sigma(\phi)$ curve. For the granular systems without adhesion between particles, jamming transition depends on the path to reach the phase boundary. For example, along the ϕ-axis, when the density of the isotropically compressed granular systems reduces to a certain point, and the volume elastic modulus is not zero, this transition point is known as point J, and the corresponding critical packing fraction is ϕ_C. $\Delta\phi = \phi - \phi_C$ is defined as the distance to point J. Granular systems show some critical behaviors near point J. For example, the volume elastic modulus shows the power scaling of $\Delta\phi$. Currently, research into the characteristics of point J remains an intense area of study in soft condensed matter physics.

As ϕ increases from 0.828 by a small step of 10^{-5}, the variation of boundary pressure P and the coordination number Z are shown in Figure 8.12(a). At the beginning, P and Z are zero, but when ϕ increases to 0.83049, P suddenly changes to 667 Pa and Z to 2.776. This point is point J, and ϕ_C is 0.83049. Figure 8.12(b) shows the change details of P and Z near point J. With the continuous increase of ϕ, P increases greatly, while Z increases slowly, i.e., Z remains at approximately 3.7.

Figure 8.13 shows the scaling behavior of P and Z near point J, i.e., $(\phi - \phi_C) < 0.01$. It can be seen that both of them obey power law scaling along with $(\phi - \phi_c)$, $(P - P_C) \sim (\phi - \phi_C)^{1.20}$, $(Z - Z_C) \sim (\phi - \phi_C)^{0.25}$. For the granular system with particles possessing smooth surfaces, the theoretical values of the power exponent are 1.0 and 0.5 for P and Z, respectively. Friction has little influence on the power exponent of the boundary pressure, but it has a greater influence on Z. There are no satisfactory physical explanations so far for the effects of friction.

The measurement circles MC1 and MC2 are placed inside and outside the shear bands to track variations of ϕ and Z, as shown in Figure 8.14.

It can be seen that the packing fractions in both MC1 and MC2 first increase slightly. When $\varepsilon = 2.8\%$ (point B), ϕ of MC2 shows a slight decline. This could be explained by the presence of some microbands. Before $\varepsilon < 4.2\%$ (point C'), the changes of ϕ in both MC1 and MC2 are very small. When $\varepsilon = 4.2\%$, ϕ of MC2 drops consistently, and the granular system enters the critical state with a fully developed shear band. At this time, ϕ of MC1 remains at

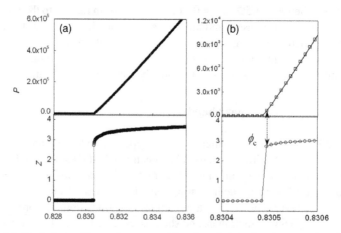

Figure 8.12: (a) Variation of boundary pressure P and coordination number Z, and (b) the determination of the critical packing fraction ϕ_C.

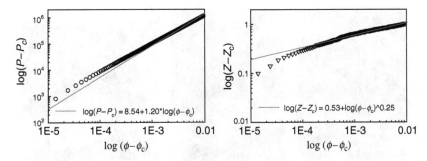

Figure 8.13: Near the jamming point J, the power law scaling of boundary pressure P and coordination number Z with $(\phi - \phi_c)$.

about 0.85. The fluctuations are slight. ϕ of MC2 continually decreases, which implies that the shear band becomes loose. Similar results for Z can be found in Figure 8.14(b).

From the evolution of ϕ of MC1, the ϕ outside the shear band is always greater than ϕ_C, which indicates that the region is jammed; ϕ of MC2 inside the shear band is always less than ϕ_C due to shear dilatation, which indicates the region is unjammed. Therefore, the phenomena of shear bands correspond to deep physical mechanisms. The jamming phase diagram analysis may be an effective way to understand shear bands, but a lot of additional work needs to be carried out.

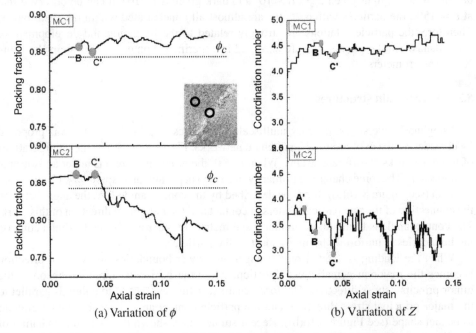

(a) Variation of ϕ (b) Variation of Z

Figure 8.14: Variation of (a) ϕ and (b) Z with the axial strain. The dotted line corresponds to ϕ_c. All marked points correspond to Figures 8.14 and 8.15.

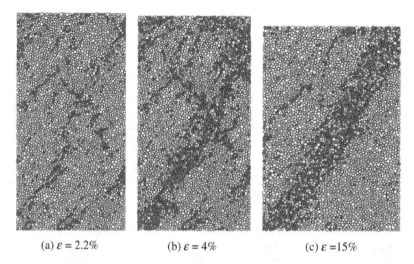

(a) $\varepsilon = 2.2\%$ (b) $\varepsilon = 4\%$ (c) $\varepsilon = 15\%$

Figure 8.15: The distribution of rotation angle α at various strains. White denotes $\alpha \leq 5°$, light green $5° < \alpha < 10°$, and dark green $\alpha \geq 10°$. (a) $\varepsilon = 2.2\%$. (b) $\varepsilon = 4\%$. (c) $\varepsilon = 15\%$.

It is necessary to track the accumulated rotation angle α from the beginning to a strain state of ε. Figure 8.15 shows the vortex-like motion of particles near the shear band. White denotes $\alpha \leq 5°$, light green $5° < \alpha < 10°$ and dark green $\alpha \geq 10°$. It can be observed that at $\varepsilon = 15\%$, the particles with $\alpha \geq 10°$ are almost all concentrated within the shear bands. Therefore, the particle rotations are strongly related to the formation and development of shear bands. Similar phenomena are observed in experiments on dry sand systems by using X-ray spectrometers.

8.3.3 Force-chain structures

At Tsinghua University, a primary multiscale mechanics of granular matter was proposed by considering the evolutions of force-chain networks. Figure 8.16 shows the configurations of force networks at different strains. When $\varepsilon = 0$, the granular system is under an isotropic compression. The force chains are primarily in a ring shape, and their spatial distributions are uniform (see Figure 8.16(a)). It can be described by the contact angle, i.e., the angle between the connection of the two particle centers in contact and the horizontal direction of this work. The contact angle is found to be circularly symmetric. In other words, the loading acting on the boundaries is uniformly borne by most of the particles.

When the loading is exerted on the upper and lower boundaries, the vertical direction becomes the major principal stress direction, while the horizontal direction becomes the minor principal stress direction. The force-chain structure evolves in to columns parallel to the major principal stress direction, and the particle contact angle changes from a circular to peanut shape (see Figure 8.16(b)). Recent studies have shown that with the changing of the direction of principal stress, the contacts on the direction of the minor principal stress decrease, which is similar to the results in this work.

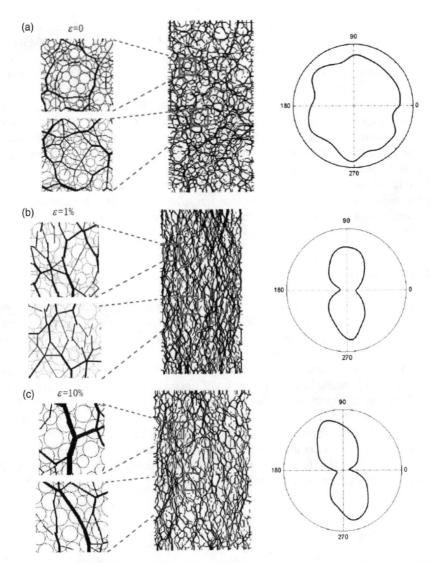

Figure 8.16: Configurations of force chains. (a) Circular force chains. (b) Columnar force chains. (c) Coexistence of circular force chains inside the shear band and columnar force chains outside the shear band.

At the stress dilatancy stage, where $\varepsilon = 2.8\%$, local pores become larger, and force chains return to a ring shape (or vortex shape) and form the so-called arches. Since the contact number is lower and the strong force chains lack support from the weak force chains, the particles exhibit an obvious rotation. Therefore, the circular force chains are very unstable and easy to rupture. The breakage of force chains is a process equivalent to the occurrence of microbands; several microbands results in the formation of shear bands.

When the granular system forms a large number of shear bands, the force chains outside the shear bands remain as columnar force chains parallel to the vertical direction. On the other hand, the force chains inside the shear bands bend, which indicates that the principal stress of the granular system within the shear bands experiences torsions (see Figure 8.16(c)). The bent force chains form circular force chains in the shear bands, and the width of circular force chains is consistent with the width of the shear bands. When a shear band is formed, the contact angle experiences certain changes at this time, and some force chains that form in the shear bands are not parallel to the principal stress direction, but are at an angle. This can lead to the situation where contacts in the bending direction comprise the largest group of contacts, and where the contact angle distribution exists as a peanut-shape.

The force-chain configurations during the application of different strains reflect the resistance strengths of granular systems to external loadings. It is necessary to conduct a detailed analysis on their structural characteristics and evolution dynamics.

8.4 Energy transformations

Granular matter consists of a large number of clearly distinguishable particles, where neighboring particles are in contact with one another and form skeletons to support gravity and other external loading forces. Some examples include coarse grains of soil and rockfill material. Once the contacts are lost or recreated, various kinds of energy, such as elastic energy or kinetic energy, are transformed into one another, while some types of energy are dissipated if sliding occurs. In this paper, a biaxial test on a two-dimensional granular sample is numerically simulated by using the software PFC2D. The fluctuation in strain–stress relations is observed and considered as the stick–slip motion of a force chain, i.e., destruction and reconstruction of a force chain. Every type of energy is calculated at different stress–strain stages. They are also compared with force-chain network configurations. It is found that as shear bands fully develop, the boundary work mostly dissipates within the shear band. These findings emphasize the importance of studying mesoscale structure dynamics, such as particle and force chain re-arrangement.

8.4.1 Introduction

Granular matter consists of individual particles. The typical size of these particles ranges from tens of micrometers to tens of meters. Some examples of such matter are encountered in coarse granular soil and debris flows. In many engineering problems, it is usually treated as a continuous medium with certain mechanical properties. However, the assumption of it being continuous develops into an obstacle against further progress as soon as stress or strain localization (shear bands) occurs around a peak stress. This is because the shear bands would cause granular matter failure. As the initial constitutive model cannot properly predict the appearance of shear bands, detailed studies have been conducted over the past few decades, and many improved models and theories have been proposed. These theories include the gradient plasticity theory, bifurcation theory, etc. In particular, the non-localized damage model considers the microscopic structure of the materials, such as the interaction between the strain gradient, and microcracks. The traditional mechanical model of localized damage is extended to the non-localized damage mechanics model, i.e., the stress of one point is not only related to its strain change history, but is also affected by the interaction of material points in a certain neighborhood.

A large number of tests and simulations have shown that the macromechanical properties of granular matter are affected by the fabric and particle properties on a microscopic scale. Oda (1997) and Oda and Kazama (1998), observed the microstructure of shear bands that developed in several natural sands, by X-ray application as well as by an optical method using a microscope and thin sections. Despite there being extensive studies on shear bands and microfabric properties, the microdeformation mechanisms leading to the development of shear bands are not yet clearly understood, and more importantly, the localization of strain remains a pressing topic in the theoretical and experimental studies on the mechanics of granular materials.

In order to further study the effect of microfabric on the mechanical properties of granular matter, an alternative to these continuous approaches is to use discrete variable-based methods that represent the material as an assemblage of independent elements (also called units, particles or grains), which interact with one another. The commonly adopted numerical method for a discrete system is the particle flow code (PFC) that is particularly suited in simulating granular materials. It immediately abandons the assumption of the continuum and considers such matter as a conglomeration of discrete particles. It begins with the contacts among particles, which obey the contact law, such as Hook's linear contact law or Hertz's nonlinear contact law. It has also been extended to solid mechanics to investigate the failure process of geomaterials. Currently, huge datasets of particle motion can be easily obtained in experiments and in discrete-element simulations, and complicated behaviors can be well understood from the perspective of the motion of particles. For example, it is clear that the behavior of shear bands is caused by the sliding and rotation of strong particles. However, it remains difficult to choose an appropriate theory to explain or successfully predict the formation and development of the shear bands in granular matter.

By observing the shearing failure, it has been found that the range of influence of shearing is only limited to a length several times the particle diameter, while the particles beyond this range are almost free of shear effects. Recent studies have emphasized the importance of force chains, which are formed by quasi-linearly aligned particles. These quasi-linear structures bear and transmit the compressive load on the system. The persistence of a force chain could resist the repulsive contact forces that push particles apart and try to break the chain. The transverse contact force, applied by particles belonging to weak chains, is the main reason that the chain breaks. The force-chain intensity is different in different layers of the granular system; the deeper the depth, the more stable the force chain is. When the granular system is evenly split along the direction of the shear bands to a thickness of n-particle layers, the force chain in each layer corresponds to a critical force-chain friction. When the applied stress is greater than the critical value, the force chain will deform, fracture or reorganize, giving rise to the damaged force-chain network. Therefore, the internal particles will slide and rotate relative to another and cause the energy losses. Antoinette [3] studied the force chains' buckling process and established the relations between stability and energy of particle samples.

High-frequency acoustic waves are often used in the abovementioned experiments, but few experiments on the frequency response under low-frequency shear have been reported [5], [7]. However, in geotechnical engineering, the shear bands always appear in the loading process under a quasi-static condition. Therefore, studies on the elastic energy, kinetic energy, and other types of energy of the entire granular system are very important for examining the deformation of granular matter experiencing biaxial compression.

In this paper, the macro- and micromechanical behavior of the granular sample under biaxial compression is simulated with the software PFC2D. The transformation of elastic energy and kinetic energy, energy dissipation, and the evolution of energy with force chains are discussed; this is an attempt to explain the mechanical properties of granular matter from both energy and mesostructure perspectives.

8.4.2 Simulation setup

In this simulation, a two-dimensional granular sample 0.2 m high and 0.1 m wide is selected. It consists of 4325 round particles. The particle size obeys a uniform probability distribution in the range of 0.0046 to 0.0075 m. The surface friction coefficient $\mu = 0.5$, normal and tangential stiffness are both 5.0×10^8 N/m. Time step is $\Delta t = 10^{-6}$ second and the gravity is set to 0. During the loading process, a constant confining pressure of 2 MPa is applied in the horizontal direction, and 0.1 mm/s stress acceleration is imposed in the vertical (i.e., axial) direction of the top boundary and bottom boundary. Figure 8.17 shows the stress–strain curve and the changes in the force chain of the granular sample. The partial stress $\sigma = \sigma_1 - \sigma_2$, in which σ_1 is the axial stress and σ_2 is the level of confining pressure. The axial strain is ε.

During the tests, the loading speed is required to be slow, so as to ensure that the test is conducted in a quasi-static condition. It is essential to ensure the quasi-static nature of the deformation process. If the loading process is in a dynamic condition rather than a quasi-static one, the force chains and movement of particles will behave very differently; this has not been mentioned in past studies on this topic. Therefore, this paper analyzes the loading process under a quasi-static condition. This can be verified by comparing the boundary work with the elastic energy in the granular sample. If both are equal in the elastic component, it satisfies the quasi-static condition. Otherwise, the loading is under a dynamic condition.

Figure 8.17 shows stress–strain curves and the evolution of force chains. One reason for the large stress–strain fluctuation is the scale effect, i.e., the number of particles (4325) is not large enough. Another major reason is the stick-slip motion of force-chain networks (Ciamarra et al. [5]). Force chains are self-adaptive under external loadings, in other words, the configuration of force chains transforms instantaneously into an appropriate stable state. If an applied force is large enough to cause larger force chain configurational changes, a sudden decrease in the stress of the granular system would be initiated, as shown in the above stress–strain relation. We can see the frequent transformations of force-chain configurations, as shown in the right inset, even within a short duration of 10^{-6} second. However, the question of how to characterize the stick-slip motion of force-chain networks remains unanswered.

The two arrows shown in Figure 8.17 from left to right, respectively, represent the starting point of the elastic and plastic stages. This can be determined by the subsequent

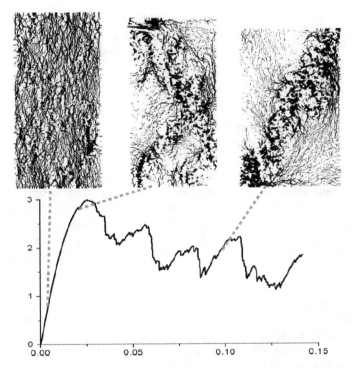

Figure 8.17: The curve of stress with strain and three diagrams of force-chain network transformation. Each diagram shows the difference of force chain during a time step of 10^{-6} second. The inset denotes the schematics of biaxial compression.

kinetic energy and friction energy pulse. From the stress–strain curve, we can see that the granular system experiences a complete elastic–plastic process. When $\varepsilon \approx 1.9\%$, the system exhibits significant elastic properties. With the increase of the axial strain, the system enters the plastic stage, and the localized shear bands emerge. Initially, these localized shear bands, so-called microbands [2], [4], are very small and are randomly distributed within the entire granular system. With further increases in the axial strain, the localized shear bands begin to deliver, combine, and form a number of distinct shear bands. Eventually, they form an X-type shear band. When $\varepsilon = 2.8\%$, the granular system achieves peak strength, with a subsequent rapid decline in stress. A single shear band appears along the direction of around 45°, after which the system enters the softening stage and the residual stress stage.

The above results can also be obtained by observing the force chain changes in Figure 8.17. At different stress–strain stages, the force-chain changes are not the same. At the elastic stage, the evolution of the force chain primarily manifests as the thickening of the force chain, and as cracks initiated by sliding that are evenly distributed in the granular system. When the system enters the plastic stage, the process of evolution of the force chains chiefly shows a large number of dispersed, short, shear-sliding force chains. When the shear bands are formed, the evolution of the force chain is primarily centralized in the shear bands, and the changes that occur manifest as the formation and destruction of the force chain in the shear direction. At the same time, the number of force chains inside the shear bands is significantly lower than outside. In addition, the force chain inside is significantly weaker than the one outside.

8.4.3 Energy analysis

In shear failure under biaxial compression, there mainly exist the following forms of energy: the elastic energy, the kinetic energy, and the frictional energy dissipation. Each of them will be analyzed, one by one, as described. Regarding energy losses, there are two major types in granular systems. One type is the energy dissipation caused by mutual collisions between particles, which is usually defined by the camping coefficient. In this paper, the collision is defined as a completely elastic collision, namely, the coefficient of restitution is 1, and the damping coefficient is 0. The other type is the energy dissipation caused by friction. One part of the particles loses contact while another part newly establishes contact under the external loading. During these changes in contact, there exists energy dissipation, which is the focus of this paper.

8.4.3.1 Elastic energy and critical sensitivity
The static loading method is used, so that at each stage, the deformation is in equilibrium. The particle contact model is elastic and the elastic energy is stored through the contact deformation of the particle. The elastic energy formula can be seen in Appendix A.

The results of Ee are shown in the inset of Figure 8.18. From the fluctuation of Ee, it can be ascertained that the granular sample might become sensitive to axial strain prior to failure, i.e., critical sensitivity [6]. Since the controlling variable is the axial strain ε in the simulations, the sensitivity of the release in elastic energy under axial strain is defined by:

$$S = \left(\frac{\Delta E_e}{\Delta \varepsilon}\right)\left(\frac{E_e}{\varepsilon}\right)^{-1} \tag{8.5}$$

where ΔE_e is the response of the release in elastic energy to axial strain increment $\Delta \varepsilon$. The sensitivity S is shown in Figure 8.18.

From the inset of Figure 8.18, it can be observed that the elastic energy curve is very similar to the stress–strain curves, indicating that when subjected to external loading, the granular system shows an elastic effect. In other words, with an increase in stress, the elastic energy also increases. The sensitivity S is shown in Figure 8.18. When $\varepsilon < 1.9\%$, S remains low and nearly a constant, indicating that in the elastic stage, the external loading is completely converted to elastic energy. This is easy to understand. When $\varepsilon = 1.9\%$, the first pulse of S occurs, after which S increases and oscillates. The system begins to undergo plastic deformation and the local force chains undergo fracture and reorganization, in combination with the subsequent pulses. Moving into the plastic stage, the granular sample dissipates energy due to plastic deformation. Since ideal elastic particles are used in this simulation, the particles themselves do not undergo plastic deformation. Therefore, in this stage, the plastic energy is the energy released by the unrecoverable structural changes of local force chains.

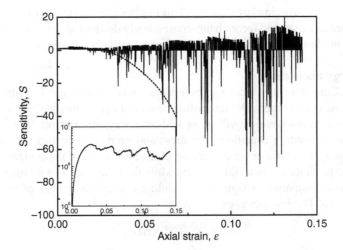

Figure 8.18: Critical sensitivity in the biaxial test, where the inset denotes the elastic energy. The dotted curve outlines the tendency of S. Initially, a few sharp drops in S imply the subsequent occurrence of shear bands.

When $\varepsilon = 2.8\%$, the elastic energy achieves the peak, namely, the maximum energy storage of elastic energy. The cyclical fluctuation is closely linked to the structural changes of the particles in the shear bands. At this stage, the sensitivity decreases very sharply ahead of the eventual appearance of shear bands. Therefore, it appears that critical sensitivity may be a reasonable indication of the occurrence of shear bands.

8.4.3.2 Kinetic energy
Kinetic energy is predominantly caused by the translation and rotation of particles during the process of breaking and restructuring of particle contacts. The variation in kinetic energy can be obtained by calculating the changes in particle velocity field; the calculation formula is:

$$E_c = \sum_{k=1}^{N} e_c^k \tag{8.6}$$

The kinetic energy variation of a single particle is:

$$e_c^k = \frac{1}{2}mv^2 + \frac{1}{2}I\omega^2 \tag{8.7}$$

where m is the particle mass, v is the translational speed, I is the moment of inertia, and ω is the rotational speed. Since the loading process in this study is under a quasi-static condition, the kinetic energy is as small as approximately 1% of the elastic energy, as shown in Figure 8.19.

At the elastic and plastic stages, the kinetic energy is close to zero, due to the quasi-static loading of the granular system. At the softening region and the residual phase, the kinetic energy shows large fluctuations. When examined with the measurement of elastic energy shown in Figure 8.18, the positions of increments in the kinetic energy correspond exactly to the positions of decreases in elastic energy. This implies that the release of elastic energy first drives particles to move, and force chains consequently destruct and reconstruct. Eventually, the decrease in elastic energy is largely transformed into plastic energy.

8.4.3.3 Energy dissipation

According to Coulomb's law, when the friction force reaches the upper limit, the particles slip and roll against one another. The interrolling does not cause microdeformation; therefore, if all particles roll, deformation will occur without energy dissipation. When the system is under quasi-static loading conditions, the stress anisotropy and contact network anisotropy in the granular system arise not only because of the anisotropy of the contact force, but also because some particles are in a sliding state, while the others are in a rotating state.

The friction dissipation is equal to the sliding energy dissipation of all contacts at a certain time step. The formula is described as:

$$E_f = \sum_{N_c} \left((\langle f_i^s \rangle)(\Delta U_i^s)^{\text{slip}} \right) \tag{8.8}$$

where N_c is the number of particle contacts, f_i^s is the average shear stress, and $(\Delta U_i^s)^{\text{slip}}$ is the incremental sliding displacement of the particle contacts at a certain time step. The incremental sliding displacement can be determined by the following formula:

$$\left(\Delta U_i^s\right)^{\text{slip}} = \Delta U_i^s - \left(\Delta U_i^s\right)^{\text{elas}} = \Delta U_i^s + \frac{\left(\Delta f_i^s\right)^{\text{elas}}}{k^s} = \Delta U_i^s + \frac{\left(f_i^s\right)^{(t+\Delta t)} - \left(f_i^s\right)^{(t)}}{k^s} \tag{8.9}$$

Here, ΔU_i^s is the incremental shear displacement, which is comprised of the elastic displacement $(\Delta U_i^s)^{\text{elas}}$ and the sliding displacement, as specified in the manual of PFC software. Figure 8.20 shows the dissipated energy due to friction sliding during the loading process. It can be observed that there is a roughly linear relationship between E_f and ε axial strain.

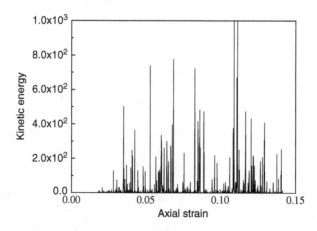

Figure 8.19: Evolution of kinetic energy.

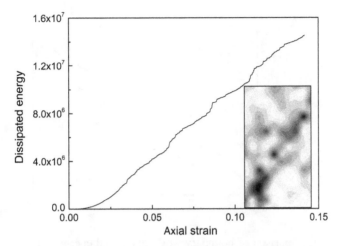

Figure 8.20: Energy dissipation under axial strain. The inset represents the contour of the number of sliding particles. The darker areas correspond to a greater number of sliding particles.

The distribution of the number of sliding particles roughly represents the distribution of energy dissipation in the granular sample. From the inset in Figure 8.20, we can observe that energy dissipation mainly occur within the shear band, where particles frequently slip. It is essential to conduct detailed studies on force structure changes and particle rearrangement.

8.4.4 Discussions

It is necessary to calculate the boundary work performed by all walls on the granular sample. The total accumulated work E_w is written as:

$$E_w = E_w - \sum_{N_w}(F_i\Delta U_i + M_3\Delta\theta_3) \tag{8.10}$$

where N_W is the number of walls, F_i and M_3 are the resultant force and moment acting on the wall at the start of the current time step, respectively, and ΔU_i and $\Delta\theta_3$ are the applied displacement and rotation occurring during the current time step, respectively. Note that this is an approximation in that it assumes that F_i and M_3 remain constant throughout the time step. Furthermore, E_w may be positive or negative, with the convention that work done by the walls on the particles is positive.

In Figure 8.21, the boundary work E_w, elastic energy E_e, and dissipated energy E_f are compared.

It can be observed from Figures 8.19 and 8.20, that with increasing axial strain when the sample is in the elastic stage ($\varepsilon < 1.9\%$), E_w is almost fully converted into E_e. At this time, the kinetic energy and friction energy account for a very small proportion of total energy. When the system enters the elastic stage ($\varepsilon < 2.8\%$), more than 85% of E_w is converted into E_e, and almost 14% is converted into E_f; only around 1% is converted into kinetic energy. After entering the softening stage and residual stage ($\varepsilon > 2.8\%$), E_w is almost fully converted

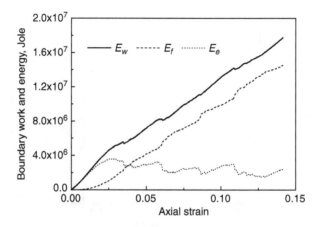

Figure 8.21: Comparison of boundary work, dissipated energy, and elastic energy.

into E_f. Further, the two slopes of E_e and E_w are very similar. This proves that the transformation of energy is concentrated in the shear bands at the residual stage, and that the particles undergo intense reorganization, rotation and other changes within the shear bands. At the same time, during the entire loading process, the proportion of kinetic energy is very small. This indicates that the translation of particles occurs on a very small scale. The analysis of the reorganization of the particles within the shear bands is discussed next.

The elastic energy release is related to the structural changes of particles in the shear bands, as shown in Figure 8.22.

Figure 8.22 is an example of a force-chain map before and after the fluctuation of the stress. When comparing the two force-chain networks, it can be observed that with the reduction in stress, some force chains undergo fractures and then disappear. The associated particles begin sliding and rotating, so the corresponding kinetic energy also increases. At the same time, new force chains are rapidly formed. The speed of formation has been presumed to be close to the speed of sound. The transition of old force chains to new force chains corresponds to the increases in stress. It is necessary to develop technology with high spatial

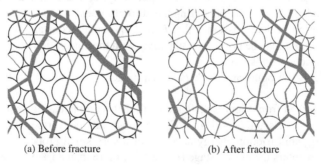

(a) Before fracture   (b) After fracture

Figure 8.22: Force chains destruct and reconstruct within a shear band. Thicker chains represent stronger forces. (a) Before fracture. (b) After fracture.

and temporal resolution, in order to conduct on-site, real-time observation and therefore capture this process.

From Figure 8.22, we also can perceive that during the hardening process up until failure, the particles are arranged in chains to form column-like structures. The directions of elongation in these structures are more or less parallel to the major principal stress axis, and the applied stress is largely transmitted through them. During the process, the pre-existing contacts are lost in the direction of the minor principal stress, but some contacts are newly formed in the direction of the major principal stress. As a result, an elongated void is generated between two neighboring columns. This is the micromechanism that leads to dilatancy before failure.

8.4.5 Outlook

Various kinds of energy are present in variable proportions at variable stress–strain stages. At both elastic and plastic stages, the boundary work is mainly converted into elastic energy. In the following residential stage, which includes the frequent slippage of particles and increases in the strength of force chains, almost all the boundary work is dissipated. More importantly, the energy dissipation is principally localized in a narrow shear band.

It should be noted that the macroshear bands are formed through a cascade: the combination and development of microshear bands, which implies a very complex process. This paper only discusses this concept from the energy point of view. Further studies will be conducted on the particles and force-chain rearrangements, by using high-speed cameras and photoelastic techniques.

Appendix A. Formulations of energies in granular systems

Under external loadings, particle motion and contacts change, corresponding to energy transformations. The following three types of energies are predominantly involved: elastic energy, kinetic energy, and friction energy.

A.1 Elastic energy

If the linear elastic contact model is adopted, the elastic energy on the contact point is:

$$E_s = \frac{1}{2} \sum_{i=1}^{N_c} \left\{ (k_n)^{-1} \left(F_i^n \right)^2 + (ks)^{-1} \left(F_i^s \right)^2 \right\} \qquad (A.1)$$

where N_C is the number of contacts between particles, F_i^n and F_i^s are the normal force and tangential force, respectively, k_n and k_t are the normal stiffness and tangential stiffness, respectively.

It should be noted that for nonlinear Hertzian contact, k_n and k_t do not appear in the Hertzian contact formula. Indeed, the dissipation of contact energy between particles is not only related to the plastic deformation, but also involves the transmission and absorption of elastic waves in the particles and on the surfaces of particles. As all these processes have a great impact on the elastic energy, another formula is required to calculate the strain energy, see details in [8].

A.2 Kinetic energy

The kinetic energy of granular system is calculated as:

$$E_k = \sum_{i=1}^{N_p} e^i \tag{A.2}$$

where N_p is the number of particles, e^i is the kinetic energy of particle i, which can be expressed as:

$$e^i = \frac{1}{2}\left(mv^2 + I\omega^2\right) \tag{A.3}$$

where m is the particle mass, v is the translational velocity, I is the moment of inertia, and ω is the rotational velocity.

A.3 Dissipated energy due to friction

According to Coulomb's law, when the friction reaches the maximum static friction, the particles begin to slip against one another. Subsequently, the energy dissipation within a time step is:

$$E_f = \sum_{N_C}\left\{\langle F_i^s\rangle\left(\Delta U_i^s\right)^{\text{slip}}\right\} \tag{A.4}$$

where N_C is the number of contacts between particles, $\langle F_i^s\rangle$ is the average tangential force, $(\Delta U_i^s)^{\text{slip}}$ is the slip displacement increment of particle contacts within a time step, determined by the following formula:

$$\left(\Delta U_i^s\right)^{\text{slip}} = \Delta U_i^s - \left(\Delta U_i^s\right)^{\text{elas}} = \Delta U_i^s + \frac{\left(\Delta f_i^s\right)^{\text{elas}}}{k^s} = \Delta U_i^s + \frac{\left(F_i^s\right)^{(t+\Delta t)} - \left(F_i^s\right)^{(t)}}{k^s} \tag{A.5}$$

where ΔU_i^s is the tangential displacement increment, $\Delta U_i^s = V^s \Delta_t$ and V^s is the tangential relative velocity. $(\Delta U_i^s)^{\text{slip}}$ is composed of two parts, the elastic displacement $\left(\Delta U_i^s\right)^{\text{elas}}$ and the slip displacement.

References

[1] C. Goldenberg and I. Goldhirsch, 'Force chains, microelasticity and macroelasticity', *Phys. Rev. Lett.*, 89, 084302 (2002).
[2] C. Goldenberg and I. Goldhirsch, 'Effects of friction and disorder on the quasistatic response of granular solids to a localized force', *Phys. Rev. E*, 77, 041303 (2008).
[3] T. Antoinette, 'Force chain buckling, unjamming transitions and shear banding in dense granular assemblies,' *Philos. Mag.*, 87(32), 4987–5016 (2007).

[4] S. J. Antony and M. R. Kuhn, 'Influence of particle shape on granular contact signatures and shear strength: new insights from simulations', *Int. J. Solids Struct.*, 41(21), 5863–5870 (2004).

[5] M. P. Ciamarra, E. Lippiello, C. Godano and L. D. Arcangelis, 'Unjamming dynamics: the micromechanics of a seismic fault model', *Phys. Rev. Lett.*, 104(23), 8001 (2010).

[6] M. R. Kuhn, 'Structured deformation in granular materials', *Mech. Adv. Mater. Structures*, 31(6), 407–429 (1999).

Chapter 9

Granular flows

Individual particles tend to move under unbalanced forces, which can be described with the Newtonian laws; macroscopically, the collection of these particles would behave like a liquid and flow like one. Roman philosopher Lucretius (98 BC to 55 BC) may have been the first one to describe the phenomenon, as he has written, "... One can scoop up poppy seeds with a ladle as easily as if they were water and, when dipping the ladle, the seeds flow in a continuous stream...". However, granular flows are fundamentally different from any other type of flow. The understanding and modeling of this common observation are rather difficult, and granular flows have been the subject of intense study at the confluence of fluid mechanics, rheology, and soft condensed matter physics. Such research is motivated by numerous applications encountered, particularly in the description and prediction of natural hazards, such as debris flows, highly fragmented rock avalanches and flows. The recent rapid progress in granular physics has greatly prompted the fundamental studies on granular flows. In this chapter, basic concepts and some new results are briefly introduced; meanwhile, the7 configurations of force chains and their evolutions in corresponding flow regimes are exhibited.

9.1 Coulomb friction

The study of granular flows can be traced back to Coulomb, who first proposed an approximate friction model:

$$F_f \leq \mu F_n \tag{9.1}$$

where F_f is the force exerted by friction (in the case of equality, the maximum possible magnitude of this force), μ is the coefficient of friction of the contacting materials, and F_n is the normal force exerted between the surfaces. The Coulomb friction law calculates the force of dry friction between relative lateral motions of two solid surfaces in contact, and provides a threshold value for this force, above which motion would commence.

In geotechnical engineering, the Mohr–Coulomb theory is used to define the shear strength of the collective particles at different effective stresses. It applies to materials for which the compressive strength far exceeds the tensile strength, expressed as follows:

$$\tau \leq c + \sigma \tan \phi \tag{9.2}$$

where τ is the shear strength, σ is the normal stress, c is the intercept of the failure envelope with the τ-axis, and ϕ is the slope of the failure envelope. The quantity c is called the cohesion and the angle ϕ is called the angle of internal friction. The Coulomb yield could be used to construct a plastic yield criterion, and only the adoption of a flow rule is required to employ the methods of metal plasticity to granular flow. As a result, it is not necessary to consider the behaviors of individual particles, but instead, the granular material is treated as a plastic solid.

In quasi-static granular flows, it is usually found that the particle concentration is always close to a critical value v_c, and it depends on external loading stress. In a great stress range, $v_c \approx 0.6$. When a granular flow is underconsolidated, the particle concentration is lower than v_c at the beginning, but gradually increases to reach v_c. Granular flows decrease to v_c when they are overconsolidated. In slow flows, granular systems can usually withstand great shear strain. In other words, the particle concentration can always be regarded as v_c, which means the granular material cannot be further compressed. Janssen has adopted this assumption when analyzing the silo effect. He proposed that beyond a certain depth from the top, the weight of particles in the silo was shared with the sidewall. The base pressure of the silo trends towards saturation, and the pressure is independent of the depth. Since the pressure controls the flow rate when particles are discharged from a silo, the flow rate is constant, which is the second most important aspect of granular flow studies.

In dense flows, particles in contact may form linear and stable force chains, and these force chains constitute the network to support the weight of particles and external loadings. If the particle concentration is low or if the particle surfaces are smooth, the collisions between particles would occur more frequently and a stable force network cannot be formed; therefore, the mechanism of force transmission would be substantially different.

9.2 Bagnold number

In order to study the bed load transport of sediment in rivers, Bagnold [1] designed an experiment to simultaneously measure the shear and normal forces generated by a sheared dispersion of neutrally buoyant particles for a wide range of solid concentrations, fluid viscosities, and shear rates (see Figure 9.1). The concentric-cylinder rheometer has an inner cylinder with a flexible membrane that allows measurements of the static pressure; the outer cylinder rotates at a constant speed and is driven by a lathe. With the lathe's traction, the outer cylinder rotates with a certain angular speed, and the friction between the outer cylinder wall and the particles leads the particles to flow in the cylinder. These finally form the granular flows, such that the particles between the inner cylinder and outer cylinder exhibit a shear movement.

To consider the interstitial fluid viscous shearing and particle collisions in granular mixtures, a dimensionless parameter N_{Bag} was defined to distinguish contributions of particle

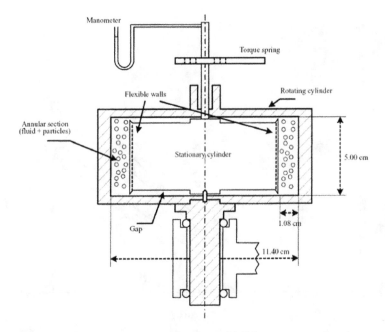

Figure 9.1: Experimental apparatus used by Bagnold [1] to measure the shear and normal forces in a sheared suspension.

collisions and viscous stress in steady uniform shear flows, as follows:

$$N_{\text{Bag}} = \frac{\sqrt{\lambda}}{\eta} \rho_{\text{p}} \dot{\gamma} D^2 \tag{9.3}$$

where λ is the linear concentration of solid particles and is usually defined as the ratio of particle diameters to their reciprocal distances: $\lambda = \left\{ \left(\frac{\nu_*}{\nu} \right)^{1/3} - 1 \right\}^{-1}$, where ν_* is the maximum value of solid concentration ν_s (for closely packed spherical balls $\frac{\pi}{3\sqrt{2}} \approx 0.74$), η is the pore fluid viscosity, ρ_{p} is the particle density, and D denotes the mean particle diameter.

Equation (9.3) indicates that N_{Bag} is not only dependent on the inertial stress of solid particles, but also on the viscous stress and volume fraction (concentration) of solid particles ν_s (commonly between 0.6 and 0.7 for granular flows).

Bagnold found that sheared layers of non-cohesive particles can experience dispersive pressure, resulting from collisional momentum transfer perpendicular to the flow direction, and stress can away from the slope. The values of shear stress τ and particle pressure P are plotted in Figure 9.2 against the corresponding shear strain rate $\gamma = dU/dy$. It can be seen from Figure 9.2 that for all values of solid concentrations within the experimental range, both τ and P become proportional to $(dU/dy)^2$ at sufficiently high speeds. This inertial regime has been further developed within a three-dimensional "kinetic" theory for dry rapid granular flows, where the bulk normal stresses in rapid, collision-dominated granular flows were also

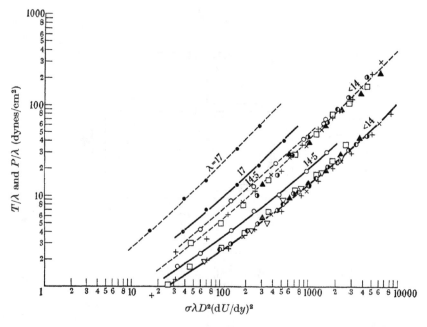

Figure 9.2: Effect of shear rate on the measured normal and shear stresses (the complete lines indicate the shear stress, and the broken lines indicate the pressure).

found to be dependent on shear rate. However, this dependence is absent in slower, friction-dominated flows. Such differences in bulk normal stress in rapid and slower flows can be ascribed to particle collisions that tend to decrease the density of the mixture by dilation of the solid phase. Bagnold [1] also showed that the relationship between bulk granular normal and shear stress was Coulomb-like, even in collision-dominated flows where shear rate $\gamma \rightarrow \infty$. An identical equation to the Coulomb model for non-cohesive materials was shown and thereby proved that Coulomb-like behavior is true for granular flows.

Furthermore, as presented by Bagnold, there exist two regimes (macroviscous and inertial), which are obtained respectively when energy dissipations occur predominantly due to interstitial fluid shear or momentum transfer via particle collisions. The analysis and experiments demonstrated that the shear and normal forces depended linearly on the shear rate in the "macroviscous" regime with $N_{Bag} < 40$ (see Figure 9.3). As the particle–particle interactions dominate in the "particle-inertia" regime when N_{Bag} exceeds approximately 200, the dominating effect of viscous forces is supplanted by inertial forces wherein momentum is transferred through particle collisions. For a collision-dominated regime, both the normal and shear stresses are proportional to the square of the shear rate and are independent of the fluid viscosity. A transitional regime falls between these two limiting values. Thus, it is clearly noted that use of the Bagnold number, the ratio of inertial stress to viscous stress, would be a rigorous means of determining the transition from viscous flows to inertial flows. The parameter N_{Bag} has since been called the Bagnold number and has been extended to dense particle flows substituting the particle density for the fluid–solid density.

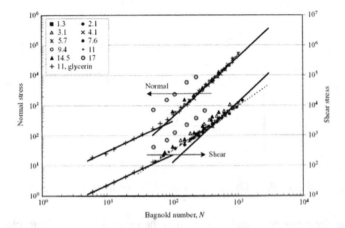

Figure 9.3: Experimental shear and normal stresses with respect to the Bagnold number for solid–water (or solid–glycerin) mixtures with different linear particle concentrations. The solid lines are the suggested correlations by Bagnold, and the dotted line is the best fit to the shear stress measurements in water.

In comparison with the continuum rheological models, the models based on Bagnold's work clearly have some limitations. Though scientifically satisfactory and widely used by geophysicists, in fundamental rheology, Bagnold's work has never been seriously confirmed and some criticism can be made regarding his interpretations of data. Bagnold's model may be relevant to describe the behavior of extremely rapid flows of dry masses of non-cohesive particles, for which it is clear that collisional effects prevail, but should be considered as mainly speculative if used for water–debris mixtures such as debris flow materials. It should be noted that the Bagnold model does not consider the possible appearance of a continuous network of interacting particles, beyond a critical solid fraction where the effect of contact friction between solid particles is dominant. At least at low shear rates, the initial Bagnold theory is unable to predict, or generate information on, debris flow yield stress, whereas this appears to be one of the basic properties of natural events. Indeed, this phenomenon of yield stress, which corresponds to a minimum shear stress that needs to be overcome for flow to take place, is behind the origin of deep debris flow deposits. Such deposits are always observed, occasionally even on steep slopes.

9.3 Inertial number and contact stress

A crucial observation raised by a da Cruz et al. [5] and Lois et al. [6] is that, in the simple sheared configuration for infinitely rigid particles, dimensional analysis strongly constrains the stress/shear rate relations. For large systems (i.e., when the distance between the plates plays no role), the system is controlled by a single dimensionless parameter called the inertial number, as follows:

$$I = \frac{\gamma d}{\sqrt{P/\rho_p}} \tag{9.4}$$

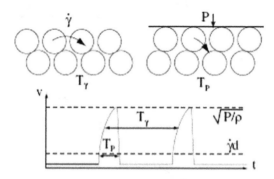

Figure 9.4: The physical meaning of inertia number I.

where γ is the shear rate of granular flows, d is the mean diameter of particles, and P is the normal stress inside granular body. ρ_p is the material density of constituent particles and d is the particle diameter.

As illustrated in Figure 9.4, a physical interpretation of the dimensionless number I is given in terms of the ratio between two time scales: (1) a microscopic time scale $\frac{d}{\sqrt{P/\rho_p}}$, which represents the time it takes for a particle to fall in a hole of size d under a certain pressure, giving the typical time scale of rearrangements; (2) a typical macroscopic time scale $1/\gamma$ linked to the mean deformation.

If Δ is the overall deformation of a chain, the deformation of each contact would be:

$$\delta = \frac{\Delta}{N} = \frac{\Delta d}{L} \tag{9.5}$$

where $N = L/d$ is the number of contacts, L is the length of a chain, and the particle diameter is d. By using Hooke's law, the contact force is given by

$$F = k\delta = \frac{k\Delta d}{L} \tag{9.6}$$

where k is the stiffness coefficient. The contact stress τ is expressed as:

$$\tau \sim \frac{F}{d^2} \sim \frac{k}{d} \tag{9.7}$$

It can be seen that the dimensionless contact stress is $\tau d/k$, and it can be interpreted as the particle deformation δ, which, in turn, is represented as a fraction of the particle diameter d.

9.4 Flow regimes

If we adopt the common definition of a "fluid" to be the state in which shear stress arises whenever there is any shear rate, then it is known that a dilute granular material is a non-Newtonian fluid with the stresses depending on the square of the shear rate. This extreme is commonly referred as the granular gas. At extremely high particle concentrations, granular materials enter the "jamming state" in which they behave solid-like where stresses no longer depend on the rate of shear. Between these two extremes, the relation between stresses and shear rate varies.

Babic et al. [3] utilized a discrete-element simulation of a two-dimensional system of a simple shear flow of monodispersed soft disks, to study the relation between stresses and shear rate. It was found that the dimensionless stress $\tau_{ij}^* = \tau_{ij}/\rho_s D^2 \dot{\gamma}^2$ depended on the dimensionless shear rate $B = \dot{\gamma}\sqrt{m/K_n}$ in such a way that $\tau_{ij}^* = a_{ij}B^{b_{ij}}$ with parameters a_{ij} and b_{ij} both being functions of the solid concentration and shear rate. In the above equation, τ_{ij} is the stress tensor, $\dot{\gamma}$ is the shear rate, ρ_s is the particle density, D its diameter, m its mass, and K_n its stiffness. A more extensive study of this stress–strain rate dependency was later conducted using a three-dimensional assembly of monodispersed spherical particles. Both the two-dimensional and three-dimensional analyses showed that at very high concentrations, the power factor b_{ij} approached -2, implying that the dimensional stress τ_{ij} became independent of the shear rate, i.e., the material approached a solid-like state. At very low concentrations, these simulation results confirmed the non-Newtonian behavior where $b_{ij} \approx 0$ and $\tau_{ij} \approx \dot{\gamma}^2$. It was also found that not only did the concentration affect the rate dependency of the stress on the strain rate, but the strain rate could also change the power factor b_{ij} in an inversely proportional manner.

Based on the simulation observations, a "regime chart" was proposed to classify the different "regimes" of a granular shear flow. This regime chart is reproduced in Figure 9.5. In this figure, "rapid flow" refers to a state where binary collisions are responsible for all stress generation, "Type A" transition refers to multiple collisions, "Type B" transition refers to force chain formations, and "Quasi-static state" refers to persistent force chains. In a phenomenological sense, one may associate the rapid flow state with the gaseous phase, the

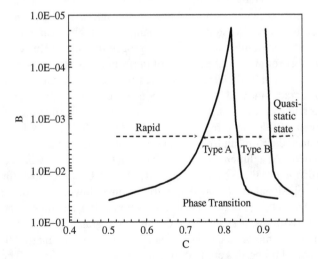

Figure 9.5: Flow regime classification (schematic regime chart, reproduced from Babic et al. [3]).

transitional states (A and B) with the liquid phase and the quasi-static state with the solid phase. The internal processes such as multiple collisions, force chain formation and its persistence, as well as the shape of the regime chart were postulated without quantitative data.

The existence of such a regime chart was also discovered in a three-dimensional study of monodispersed systems of soft spheres. A large amount of simulation data from a simple shear flow produced a "flow map" of the same shape shown in Figure 9.5. In addition, the stress-generation mechanisms leading to such transitions were also investigated, resulting in a different terminology for the various phases (inertia-collisional, inertial-non-collisional, elastic-inertial, and elastic-quasi-static). The contact time duration was determined and compared with the progression from gas-like to solid-like behaviors. It was found that the contact time dependencies on the concentration and shear rate closely resembled those of the dimensionless stresses.

The parameters in the phase diagram, such as in Figure 9.5, indicate the solid concentration and the dimensionless shear rate. These parameters are very different from the familiar thermodynamic parameters: temperature and pressure. Yet, the shape of such a diagram is strikingly similar to the phase diagram of ordinary materials. We thus ask the question: what are the internal parameters in a granular material that can help us develop an equivalent thermodynamic phase transition theory?

The contact duration and multiple collision group size were suggested as important internal parameters. This idea was investigated by Shen and Sankaran [4], where the stress, shear rate, contact-time duration, size of the multiple collision groups, as well as the coordination number were studied for a two-dimensional assembly of a soft monodispersed system. It was indeed found that the multiple collision size grew with concentration and shear rate. Though not directly measured, it is likely that when the group size grew to the size of the system, force chains spanning the system formed. Quantification of the force chain length, density, and their lifetime was not investigated due to the lack of mathematical definitions for these parameters. In addition to the internal length and time scales, the coordination number was also studied. Interestingly, at high shear rates, for any concentration, the coordination number appeared to approach the same value corresponding to the coordination number for an isostatic system of uniform disks. Recently, a three-dimensional discrete-element simulation of identical spheres in a simple shear flow was conducted to study the mean contact time and the coordination number. These parameters were studied in relation to shear rate and solid concentration for a particular granular material with a contact friction coefficient of 0.5 and a restitution coefficient of 0.7. It was found that these two parameters, contact duration and coordination number, appeared to capture the onset of transitional behavior. At a dimensionless mean contact time of ~ 2, where the dimensionless contact time is defined as the contact time divided by the binary contact time, phase change between dilute collisional flow and dense transitional flow occurred. At a coordination number of ~ 4, phase change between dense transitional flow and quasi-static flow occurred.

Based on the existence of stable force chains in the granular flows, Campbell classified granular flows into two major categories, namely, elastic flow and inertial flow. In the elastic flows, the mechanical properties are mainly dominated by force chains. The elastic flow can be further divided into elastic-quasi-static flow (i.e., the quasi-static flow in the previous classification), and elastic-inertial flow; this classification mainly depended on whether the shear stress was obviously related to the shear rate. The stress in the quasi-static flow is related to the shear strain, while the elastic-inertial flow is proportional to the shear rate. Both the flows have the same physical mechanism, relying on the force chain deformation to transmit the internal stress, and it is difficult to strictly distinguish them.

When a granular assembly is sheared with a rate γ, particles are squeezed, forming force chains in the process. The rate of chain formation is proportional to γ. The chain then rotates, is compressed, and eventually become unstable and fractures. Therefore, the lifetime of the chain is proportional to $1/\gamma$. Consequently, the product of (formation rate) $\times$ (lifetime) for a force chain is γ-independent, and the contact stress in the elastic-quasi-static flow has no relation with γ, which is exactly in agreement with the definition of quasi-static flow in the previous classification. In an elastic-quasi-static flow, particles suffer a high shear rate, and the collisions between particles are more frequent. Following this, more energy is dissipated. In order to maintain the movement of particles, a larger additional external loading should be imposed ($\sim \gamma$), so that the stress in the elastic-inertial flow is not only related to the particle elasticity, but also to the shear rate, as shown:

$$\tau = a + b\gamma \tag{9.8}$$

where a is the contact stress and $b\gamma$ is the stress due to particle inertia.

In inertial flows, collisions occur frequently; the momentum transmission between particles is only realized by collisions with each other, and therefore, stable force chains cannot be formed. The stress is the so-called Bagnold stress,

$$\tau = \rho_p d^2 \gamma^2 \tag{9.9}$$

where ρ_p is the constituent particle density and τ is proportional to γ^2. The first γ represents the magnitude of momentum exchange between particles, and the second γ reflects the collision frequency. Therefore, Bagnold stress reflects the momentum transmission caused by collisions between particles. Depending on whether the collision between particles is a binary collision or not, the inertial flow can be further divided into inertia-non-collision flow and inertia-collision flow (i.e., the rapid flow in the previous classification). For an inertia-collision flow, the collision between particles is binary, $t_c / T_{bc} = 1$, where T_{bc} is the time interval of a binary collision, and the average time interval of collisions in granular flows is t_c.

If the stiffness coefficient of a particle is constant, then T_{bc} is fixed. In order to induce a longer contact between particles, there are usually two methods. One method is that a particle is surrounded by a greater number of neighboring particles, such that it cannot be easily separated (which corresponds to the following inertia-non-collision flow); another is to lock particles in force chains, making the adjacent particles contact each other and these cannot be separated until the force chains break (the lifetime is $\sim 1/\gamma$, which corresponds to the elastic flow). Therefore, a time parameter is required to characterize the granular flow, which is the average time interval t_c of collisions. If $t_c / T_{bc} = 1$, the collision is a binary collision, which corresponds to the rapid flow; if $t_c / T_{bc} \sim 1/\gamma \sim [k/(\rho_p d^3 \gamma^2)]^{1/2}$, then the granular flow is dominated by force chains, which denotes the elastic flow. It can be seen that when the inertial flow $t_c/T_{bc} \neq 1$, the flow is neither a collision flow, nor a rapid flow. In fact, since many particles collide at the same time in the inertial flow, it is necessary to further divide the inertial flow into the inertia-non-collision flow and inertia-collision flow (i.e., rapid flow).

The quasi-static flow and the rapid flow are two exceptions. They are treated as continua in the early stage. The quasi-static flow has a higher particle concentration ($v > 0.58$), which can be described with the friction plastic model based on the Mohr–Coulomb criterion. The particle collisions in rapid flows are very similar to the molecular collisions, but energy losses occur in collisions. The gas molecular kinetic theory describing the characteristics of molecular motion is mature, which provides the basis for establishing the granular kinetic theory. It is a microscopic theory that focuses on the velocity distribution of particles; subsequently, some macroscopic properties are obtained, such as the stress, concentration distribution, energy distribution, and others. Therefore, the kinetic theory is the bridge to connect mesoscopic motion with macroscopic parameters.

Campbell [7] has questioned the basic assumption of kinetic theory – the concept of granular temperature. First, rapid granular flow with a high shear rate $\gamma \sim 10^2$ per second can only be obtained by high-speed shear experiments in the laboratory, but such shear rates are impossible to find in the granular flows in nature under normal gravity. Examples include granular flow on conveyor belts and in silos. It would be argued that under microgravity conditions, such as on the moon or Mars, a small shear force can achieve rapid flow, but with the decrease in gravity, the driving force of the granular system is reduced, and the shear rate is linearly reduced; therefore, it is also difficult to obtain rapid granular flows under a microgravity environment. Second, due to the energy consumption caused by the collisions between particles, the granular temperature must be replenished from the average flow field, which is obtained by external shear. Granular temperature controls various transport properties of granular flows, particularly pressure, and all internal/normal stresses are proportional to the granular temperature. However, the granular temperature is caused by the shear, which, in turn, follows the same order of magnitude as $(\gamma d)^2$. If all the transport rates are described by the same temperature and shear rate, the rapid flow theory cannot justify itself. Furthermore, just as the kinetic theory of gases can deduce the fluid equations, the kinetic theory predicts the adhesion response of granular matter, which does not reflect any solid characteristics (such as the supporting role of side wall friction to the gravity, a theory proposed by Janssen on the interpretation of the silo effect) of granular matter. Therefore, the solid behavior and flow characteristics of granular matter must be separated. Campbell [7] deems that the biggest problem of the granular flow theory is the introduction of granular temperature; the estimation of the granular temperature and the limited experimental data illustrate that rapid granular flow exists only in the high-speed shear experiments and computer simulations without considering the acceleration due to gravity in nature.

Since 1978, when Ogawa proposed the concept of granular temperature, there has been no work conducted to verify the reasonability of the basic assumption of the rapid flow model. The Monte Carlo simulation is based on the basic assumptions of rapid granular flow; the obtained results cannot turn to explore the reliability of rapid granular flow. The hard sphere model adopts the chaos hypothesis that is similar to the molecular collision: binary collision, which occurs in the moment and has nothing to do with history. In essence, this theory is in line with the assumptions of the rapid granular flow theory, so the results cannot be used to describe the validity of the rapid granular flow theory. At present, the rapid flow theory is looking for suitable applications, but an increasing number of granular flow systems in nature are being proven to not be rapid flows.

At the same time, Campbell has also questioned the friction-plasticity theory of quasi-static flow. For the quasi-static flow, based on the Mohr–Coulomb yield conditions of friction, the friction-plasticity theory has been established, and usually assumes that the internal friction angle ϕ between particles is kept constant (tanϕ is the apparent friction coefficient of a granular system). The theory also assumes that the particles cannot be compressed under the critical concentration state. The key concern is how to consider the boundary conditions and other mathematical problems. The friction-plasticity model has been used to simulate silo flows as well. The results are similar to the experimental results. The difference should arise from the implementation of the boundary conditions. The simulation results assume the friction angle ϕ is a constant, but ϕ is indeed variable, and why it is in a state of flux is an open question.

In addition, the elastic parameters of particles cannot be introduced into the friction-plasticity model and rapid flow theory, because these two models are based on metal–plastic and gas kinetic theory, both of which do not need to introduce the concept of elasticity of particles. For the friction-plasticity model of the quasi-static flow, only the yield boundary and flow rules are required. In the rapid flow theory, the contact between particles is considered to be instantaneous. Further, then rigid spheres (the stiffness coefficient is instantaneously infinite) are required, and no limited elasticity is allowed. Therefore, it is difficult to introduce the elastic parameters of particles in the quasi-static flow model and in the rapid granular flow model, i.e., the continuum theory of granular flow encounters almost insurmountable difficulties.

9.5 Constant-volume granular flows

Based on the existence of stable force chains, Campbell divides the granular flows into two major categories, namely, elastic flow and inertial flow. With the introduction of force chains, the complex behaviors of granular flows are easily understood. For constant-volume granular flows, the regimes are shown in Figure 9.6.

In Figure 9.6, the vertical axis is the particle concentration. The horizontal axis is k^*. It is the ratio of the contact stress with Bagnold stress, defined as:

$$k^* = \frac{(k/d)}{\rho_p d^2 \gamma^2} = \frac{k}{\rho_p d^3 \gamma^2} \qquad (9.10)$$

where k/d is the elastic stress and $\rho_p d^2 \gamma^2$ is the Bagnold stress. When k^* is greater, the elastic contact stress is dominant; when k^* is smaller, the collision inertial stress is dominant.

As shown in Figure 9.7(a), the particle concentration is high, and the shear rate is low; therefore, formation–destruction of force chains is dominant, classifying it into the elastic-quasi-static type of flows. In Figure 9.7(b), the particle concentration decreases, the shear rate increases, collisions are dominant, and only a small number of force chains are formed. This classifies it into the inertial-non-collision type of flows.

When $v > 0.59$, the flow is very clearly divided into two flow regimes: the elastic-quasi-static flow and the elastic-inertial flow. In the elastic-quasi-static flow, the stress is independent of shear rate γ, as shown in the horizontal axis $k/(\rho_p d^3 \gamma^2)$, while the

Table 9.1: Granular flow classification based on force chains according to Campell (2006)

Flow regime	Mechanism	Stress	Basic parameters
Elastic flow	(1) Particle contact deformation results in a contact force. (2) Particles in contact constitute force chains. (3) Force chains constitute a network to support the weight of a granular assembly and external loadings. (4) The particle friction coefficient has a greater effect on the formation and strength of force chains. If the shear rate is small and the friction coefficient is extremely small, no force chain forms. The flow would then transit to the inertial regime. (5) Restitution coefficient has little effect on force chains. (6) $t_c/T_{bc} \sim 1/\gamma$	Quasi-static flow (1) Particle concentration ≈ 0.60 (2) $k/(\rho_p d^3 \gamma^2)$ is greater, and the particle movement is not severe. (3) The lifetime of force chain ($\sim 1/\gamma$) × generation rate is a constant. Contact stress is independent of γ. τ, $\sigma \tan\phi$ Elastic-inertial flow (1) $k/(\rho_p d^3 \gamma^2)$ is smaller, the particle movement is severe. (2) The stress is linearly related with γ: $\tau = a + b\gamma$	(1) Contact stress caused by force chain deformation: $\tau \sim d/k$ (2) The transmission of Bagnold stress, which is a representation of momentum when particles collide: $\tau \sim \rho_p d^2 \gamma^2$ (3) The ratio of contact stress with Bagnold stress is $k/(\rho_p d^3 \gamma)$. (4) γ is the shear rate. (5) Contact forces between particles are often larger. (6) T_{bc} is the free movement time of binary collision. (7) t_c is the average of free movement time of collision.
Inertial flow	(1) The instantaneous collision between particles cannot form force chains. (2) When the shear rate is maximum, there may form force chains since the particles are carried and extruded, transiting to the elastic flow.	Non-collision flow (1) $t_c/T_{bc} \neq 1$, multiple collisions (2) $\tau \sim \rho_p d^2 \gamma^2$ Collision flow (rapid flow) (1) $t_c/T_{bc} = 1$, binary collision (2) $\tau \sim \rho_p d^2 \gamma^2$	

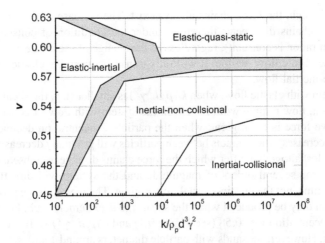

Figure 9.6: Flow regimes in constant-volume granular flows. The particle friction coefficient $\mu = 0.5$. The transition between regimes is shown in light gray [7].

elastic-inertial flow is related to γ. When γ increases, $k/(\rho_p d^3 \gamma^2)$ decreases, and the elastic-quasi-static flow transits to the elastic-inertial flow. The energy dissipation caused by inelastic collisions between particles is directly related to the restitution coefficient. It holds true only when $k/(\rho_p d^3 \gamma^2)$ is small or the shear rate is great, which makes this case significantly different from the rapid flow because the granular temperature intensely depends on the energy dissipation rate, and is also greatly influenced by the restitution coefficient. Normally, when the shear rate is low, the granular flow is quasi-static; when the shear rate increases, the system can eventually achieve the state of rapid granular flow. Assuming the particle concentration is high and constant, by gradually increasing the shear rate (i.e., decreasing $k/(\rho_p d^3 \gamma^2)$), the elastic-quasi-static flow would eventually evolve into the elastic-inertial flow, and not be involved in any inertial flows (which occurs only when the particle concentration is very low). This is meaningful, because when the particle

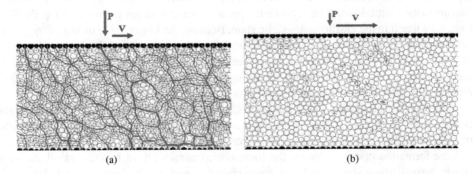

Figure 9.7: Force chains in a two-dimensional plane shear granular flow.

concentration is high, the force chains always exist. The increase of the shear rate does not make the force chains disappear; therefore, under the condition of constant volume (i.e., particle concentration is constant), regardless of the change of shear rate, the system will not depart from the elastic flow regime. It will only transit between the elastic-quasi-static flow and the elastic-inertial flow.

In the system with elastic flow, when $k/(\rho_p d^3 \gamma^2)$ is smaller (i.e., the shear rate is greater), an elastic-inertial flow regime would exist. In the system with flow of constant volume, the elastic repulsive force is often great; when the particle concentration decreases (sometimes even slightly decreases), the contacts between particles will gradually decrease and eventually contact will be lost altogether, after which the force chains disappear. Subsequently, the stress will suddenly drop several orders of magnitude and the system will directly transit to the inertial flow regime, not the elastic-inertial flow regime. The inertial force proportional to the elastic force can only be produced when the shear rate is extremely high. For example, when the particle concentration $v \approx 0.58$ (see Figure 9.1) and $k/(\rho_p d^3 \gamma^2) < 10^4$, the elastic-inertia flow will result. However, for sands with particle diameters around 1 mm, $k \approx 1.5 \times 10^4$ N/m, γ is 740 per second, and $k/(\rho_p d^3 \gamma^2) = 10^4$. In this case, the elastic-inertia flow is the likely result. In fact, such a high shear rate is unrealistic; this also demonstrates that the elastic - inertial flow of sands is almost impossible given a constant volume.

When $v \approx 0.58$, the small differences in particle concentration (such as $k/(\rho_p d^3 \gamma^2) > 1 \times 10^4$) would lead the flow regime into suddenly changing from the elastic flow to the inertial flow, which, in turn, is due to the loss of contacts between particles. In this case, the force chains cease to exist. The sudden drop in stress indicates that the elastic force chains bear a large proportion of the contact forces.

When $v < 0.56$, given a constant particle concentration, reducing $k/(\rho_p d^3 \gamma^2)$ by simply increasing γ would lead the regime into transiting from an inertia-collision flow to an elastic-inertial flow. This transition reflects a phenomenon discussed above: when the shear rate is very high, particles will be mandatorily squeezed together. Despite the large elastic-repulsive force, and the fact that the particles in contact separate quickly and do not easily form force chains, the high-speed shear forces every particle to obtain increased speed and energy in order to constitute force chains. Given a larger particle concentration, the force chains are easily formed, mainly due to the constraints of limited space. However, when lower particle concentrations and rapid shear occur, particles are compressed and balance with the elastic repulsive force to construct a few sparse and transient force chains, after which the system transits to the elastic-inertial flow. Since the particle inertia is larger in this scenario, the flow cannot evolve into the elastic-quasi-static state. Because the larger value of $k/(\rho_p d^3 \gamma^2)$ (i.e., smaller shear rate) corresponds to inertial flow, the rapid granular flows should actually be the slowest type of flow. In the inertial flows, force chains cannot form at all, and the momentum transfer is through binary collisions. Accordingly, Bagnold stress $\rho_p d^2 \gamma^2$ in the inertial flow plays a role, and is not related to the stiffness coefficient k.

Although surface friction of the particles is not the main factor that contributes to the contact stress in the elastic flow, it affects the formation and strength of force chains. If the particle surfaces are smooth, despite the occurrence of compression in the normal direction and the formation of force chains, the force chains cannot afford to shear at all. In other words, a small shear stress can break force chains; therefore, the friction coefficient has a greater impact on the force chains, subsequently affecting the granular flows. For example, when the particle concentration is 0.6, the shear rate γ is high (i.e., $k/(\rho_p d^3 \gamma^2)$ is small), and the friction coefficient decreases from 0.5 to 0.1, this leads to the result that force chains cannot form. In this case, the system with granular flow transits from elastic flow to inertial flow, and the stress transits from the contact stress to the Bagnold stress, decreased by about two orders of magnitude.

9.6 Constant-stress granular flows

The particle concentrations of most granular flows in nature, such as debris flows, do not remain the same; they all have free surfaces. The free surfaces tend to expand or contract in order to balance the external loading, thus the stress in these flows is controlled by the external loading. Sometimes, the particle concentration may experience small changes, but these changes have a significant impact on the rheological properties of granular flow. Therefore, the behavior of such systems is quite different from those with constant stress and constant particle concentrations. The latter systems can be classified as being similar to the studies of systems with entirely different shear behavior, such as soil mechanics of soil with saturated water under drainage or non-drainage conditions. Under non-drainage conditions, the fact that the soil is saturated with water, and given the assumption that the water as incompressible, the water filling the pores between particles will prevent the soil volume from changing (which corresponds to the constant volume of granular flow). In such a system, the external loading must balance the contact force within the particles in the soil and under water pressure. On the other hand, under drainage conditions, the remaining force within the soil is only the acting force between particles (which corresponds to the constant pressure of granular flow); the assembly can resist the external loadings only through the elasticity or inertia between particles; therefore, at a small shear rate, even a small external loading is imposed on the granular system. As contacts occur between particles, such particles constitute force chains, giving rise to elastic flow. Therefore, there could be a smooth transition between these flows. As for the system with a constant volume, the situation is different. Occasionally, despite the concentration increasing by only 1%, the contacts between particles are completely lost, resulting in the decrease of stress by several orders of magnitude.

The vertical axis in Figure 9.8 is $\tau_0 d/k$, where the external loading stress is τ_0. Compared with Figure 9.7, it can be found that there is a big difference between the two. It needs to be noted that the collision flow occurs at a small value of $k/(\rho_p d^3 \gamma^2)$ rather than a larger value of $k/(\rho_p d^3 \gamma^2)$ in the constant volume system, which indicates that the increase in shear rate can direct the continuous transition from the elastic-quasi-static flow to the elastic-inertial flow, and subsequently, the inertial flow. This transition order is almost diametrically opposed to the constant volume granular system. There is no regime of inertial-non-collision flow in Figure 9.8, because the transition from the elastic-non-inertial flow to the inertial-collision flow is very quick. Furthermore, this region is very narrow and can only be observed under certain circumstances.

In Figure 9.8, it is observed that the stiffness coefficient of particles is almost independent of the shear rate. For some soft materials, k is smaller, and $k/(\rho_p d^3 \gamma^2)$ is reduced, which forces the granular system to transit to the inertial flow. At the same time, $\tau_0 d/k$ is increased,

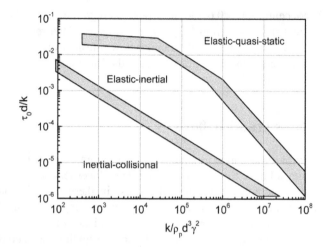

Figure 9.8:   Flow regimes in constant-stress granular flows. The particle concentration is 0.7 and the particle friction coefficient $\mu = 0.5$. The transition between regimes is shown in light gray [7].

which diverts the system away from the inertial flow. Therefore, the change in k has little impact on the granular flows where constant stress is involved.

9.7 Regime transition

The intensity, duration, and spatial distribution of force chains are the chief mechanical properties to describe the granular flow transition. By analyzing the instability, fracture, and reconstruction of force-chain structures, the macrostatic mechanical behavior and rheological characteristics of granular matter can be thoroughly understood, and then the critical parameters of the flow transition can be determined. In the discrete-element simulations of granular phase transition, the simple shear flow is the most common type of flow. Ji and Shen simulated the simple shear of polydispersed particles. The maximum and minimum particle sizes are $1.1D$ and $0.9D$, respectively, where D is the average particle diameter, and the particle sizes are distributed by equal probability in this interval, as shown in Figure 9.9. The linear viscous-elastic contact force model and the periodic boundary conditions are adopted.

9.7.1 Macrostress

The dynamic processes of granular flow under shear stress are simulated at different concentrations and shear rates, and the simulated dimensionless macrostress $\sigma_{ij}^* = \sigma_{ij}/\rho D^2 \gamma^2$ is plotted in Figure 9.10. Here, the shear stress σ_{12}^* is in the x–y plate, the normal stress σ_{22}^* is in the y-direction, and the dimensionless shear rate is $B = \gamma \sqrt{\rho D^3/K_n}$. The other stress components σ_{11}^* and σ_{33}^* have similar distributions to σ_{12}^* and σ_{22}^*. If the granular flow is divided, based on its dynamic behavior, into three states of motion, namely, rapid flow, slow

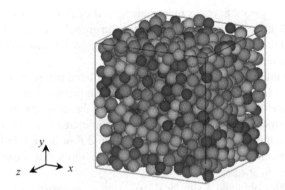

Figure 9.9: Initial particles of simple shear flows. Colors denote the particle size.

flow, and quasi-steady state, the flow properties and the quasi-solid–liquid phase transition can be obtained from the simulated results.

From Figure 9.10(a) and (b), it can be seen that the dimensionless macrostress σ_{ij}^* (= $\sigma_{ij}/\rho D^2 \gamma^2$) is independent of the dimensionless shear rate $B = \gamma\sqrt{\rho D^3/K_n}$ at low particle concentrations ($C = 0.40$). It also means the macrostress σ_{ij} does not have any relationship with material stiffness K_n, but is proportional to the square of the shear rate γ. In this situation, the granular material appears to take the form of rapid flow, and has fluid flow properties. At high concentrations ($C > 0.60$), σ_{ij}^* represents the linear function of B with an exponential slope of -2. This means that the macrostress σ_{ij} is independent of the shear rate γ, but has a linear relationship with stiffness K_n. In this situation, the granular material lies in the quasi-steady state, and has the properties of solid material. At intermediate

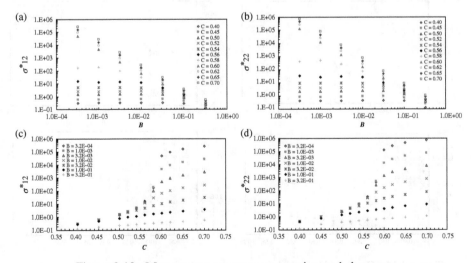

Figure 9.10: Macrostress versus concentration and shear rate.

concentrations, i.e., $C = 0.58$, σ_{ij}^{*} is independent of B at slow shear rates, and has a linear relationship with B at fast shear rates. This means the granular material changes from quasi-static flow to rapid flow when B is decreases, and the granular material experiences a phase transition from quasi-solid to liquid.

To analyze the influence of concentration, the simulated macrostress is plotted in Figure 9.10(c) and (d) with C as the x-coordinate. At high shear rates, σ_{ij}^{*} increases slowly with the increase in shear rate. With the decrease in shear rate B, the influence of concentration on σ_{ij}^{*} becomes more and more obvious; σ_{ij}^{*} exhibits a drastic increase at $C = 0.58$. Therefore, the granular material manifests as quasi-static flow or rapid flow at high or low concentrations, respectively, and manifests in various flow statuses at intermediate concentrations, depending on the variable shear rates.

Shen et al. (2004) obtained a similar distribution of macrostress with two-dimensional and three-dimensional discrete-element method (DEM) simulation, respectively, and also discussed phase transition from a rapid flow dominated by inertial collisions to a quasi-steady state dominated by elastic collisions. In the following sections, the parameter characteristics in the phase transition will be discussed, based on the simulated contact time number and coordination number in different phases.

9.7.2 Contact time number

The contact time number m and the net contact time number m' are plotted in Figure 9.11. From Figure 9.11(a) and (b), it can be seen that the net contact time number approaches zero, i.e., $m' \rightarrow 0$, at low concentrations and low shear rates. With the increase in shear rate, the net contact time number increases and approaches unity, i.e., $m' \rightarrow 1$. The contact time number has the largest value, and even exceeds 10^3 at high concentrations and low shear rates. With

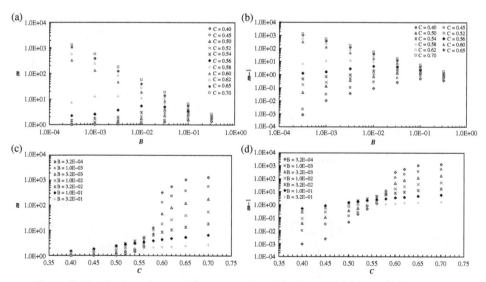

Figure 9.11: Contact time number versus particle concentration and shear rates.

the increase in shear rate, the net contact time number decreases, and we also have $m' \rightarrow 1$. Therefore, the contact time can be sustained longer, the force chains barely break, and the granular material appears to take the form of quasi-static flow at high concentrations and low shear rates. Most of the collisions are binary collisions, and the force chains break easily without strong exterior resistance at low concentrations and low shear rates. Under rapid flow conditions, the contact time number has a weaker relationship with concentration, and the granular material lies in a dynamic state with high-frequency random collisions.

From Figure 9.11(c) and (d), an interesting phenomenon can be observed. The contact time number increases with the particle concentration, and has an intersection with the net contact time number of $m' = 1$ at an approximate concentration $C = 0.56$. This is because of the variation in the rate of increase and amplitude at different shear rates. If $C < 0.56$, the contact time increases with an increase in shear rate; if $C > 0.56$, the contact time increases with a decrease in shear rate. Therefore, the concentration is a principal factor affecting the generation and breakup of force chains, and even the force transition and distribution. Therefore, we would like to introduce this critical concentration $C^* = 0.56$ as a characteristic number to determine the phase transition. Around this concentration, the contact time number is independent of shear rate and is maintained as a constant number $m^* = 2.0$. When $C > C^*$, we obtain $m > m^*$, and the granular material appears as a quasi-static flow; When $C < C^*$, we obtain $m < m^*$, and the granular material appears as a rapid flow. When C is around C^*, we obtain $m \rightarrow m^*$, and the granular flow undergoes a phase transition at different shear rates. Therefore, the critical contact time number m^* can be used as a characteristic parameter to determine the phase transition of granular materials.

9.7.3 Coordination number

The simulated coordination number versus concentration and shear rate is shown in Figure 9.12. From Figure 9.12(a), we can find that the coordination number increases with the increase in shear rate at low concentrations, and decreases with the increase in shear rate at high concentrations. Thus, we can speculate that the coordination number can approach zero, and the granular material appears to be free flowing at low concentrations and slow shear rates. When the coordination number increases further at high concentrations and slow shear rates, the granular material becomes packed more densely, and appears to take the form of quasi-static flow. Under conditions of rapid flow, the coordination number has little relation with concentration. From Figure 9.12(b), we can also see that the coordination number increases with the increase in concentration, but the rate of increase is faster at slow shear rates.

The distribution of the coordination number at different concentrations and shear rates is also similar to that of the contact time number. The coordination number decreases with the increase in shear rate at high concentrations, and with the decrease in shear rate at low concentrations. At an intermediate concentration ($C = 0.59$), the coordination number is independent of shear rate, and remains at around a constant value of $n^* = 4.0$. The critical numbers are 4 and 6 for rough and smooth granular materials, respectively. In the current simulation of rough granular material, the friction coefficient μ is 0.5, and the simulated critical coordination number is 4; this is consistent with the analytical number. Thereby, we would like to introduce this critical coordination number n^* into the study of the phase

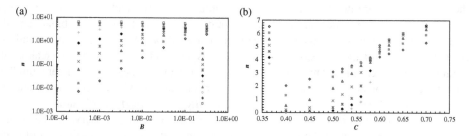

Figure 9.12: Coordination number versus concentration and shear rate.

transition of granular material. If $n < n*$, the granular material possesses a low concentration and a rapid shear rate, and appears as a liquid flow. If $n > n*$, the granular material has a high concentration, lies in the quasi-static state, and appears as a solid material. If n is around $n*$, the granular material has an intermediate concentration $C* = 0.59$, and the quasi-solid–liquid phase transition occurs at a different shear rate.

9.8 Constitutive relations

The important research considerations in granular flows are the structures of force chains in microscale and constitutive relations. The structure of force chains has been found to be quite different between granular solids and granular liquids. Current studies are primarily limited to the systems with ordinary dry particles. Despite this, a unified or widely accepted theory of constitutive relations has not been proposed.

Inspired by the viscoplastic Bingham fluid behavior, Forterre and Pouliquen (2008) proposed a constitutive equation to describe dense granular flow, which could well reproduce complex flows under different boundary conditions, as well as obtain important parameters such as the velocity distribution and particle concentration distribution. Small values of I correspond to a quasi-static regime in the sense that macroscopic deformation is slow compared to microscopic rearrangement, whereas large values of I correspond to rapid flows. The dimensional analysis illustrates that, to switch from a quasi-static regime to an inertial regime, one can either increase the shear rate or decrease the pressure. As a consequence, for rigid particles, the shear stress is proportional to the pressure, with the effective friction coefficient and the volume fraction being functions of I, as shown:

$$\tau = P\mu(I) \tag{9.11}$$

To verify the new contact friction law of equation (9.11), numerical and physical modeling of different granular flows is conducted and the test results are analyzed. Figure 9.13(a) is the test result for two-dimensional granular flows. The solid circles represent plane-shear simulations at constant pressure; particle coefficient of restitution is $e = 0.5$ and interparticle friction coefficient $\mu = 0.4$. Open circles represent rotating drum simulations in which the local friction coefficient and the local I parameter are obtained from velocity profiles at different rotation rates. The plane-shear simulation at a constant volume

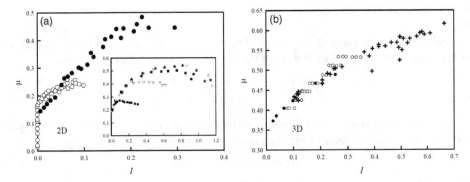

Figure 9.13: Bulk friction coefficients as a function of the inertial number: (a) data obtained on two-dimensional flows; (b) data obtained on three-dimensional flows.

fraction is also studied and the results are shown in the panel inset (using granular materials possessing the same interparticle friction coefficient $\mu = 0$ and different particle coefficients of restitution $e = 0.75, 0.5, 0.25, 0.1, 0$). Figure 9.13(b) shows the data obtained on three-dimensional flows. Open circles represent inclined-plane experiments, where μ is derived from measurements of depth-averaged velocities at different inclinations and thicknesses; solid circles represent inclined-plane simulations; crosses represent plane-shear experiments, in which normal stress and volume fraction measurements were obtained in three-dimensional annular shear cells. The figure clearly illustrates that the friction coefficient $\mu(I)$ is a function of the inertial number I. The friction coefficient is non-zero for $I = 0$, increases with I, appears to saturate at higher inertial numbers and eventually decreases when reaching the kinetic gas regime.

Based on the dimensional arguments and numerical simulations, a new constitutive law for granular flows of rigid particles can be developed, showing that the coefficient of friction on a slope bed for granular materials is dependent on the contact normal stress and shear rate, as shown:

$$\mu(I) = \mu_s + \frac{\mu_2 - \mu_s}{(I_0/I) + 1} \tag{9.12}$$

where μ_s, μ_2, and I_0 are granular material constants. It can be seen from Figure 9.13 that for quasi-static granular flow, the force chains last for a long time; when $I_0/I \rightarrow \infty$, $\mu(I) = \mu_s$. For rapid flow, the force chains last for a very short time; when $I_0/I \rightarrow 0$, $\mu(I) = \mu_2$. The particle concentration ν is the function of the inertia number I:

$$\nu = \nu_{max} - (\nu_{max} - \nu_{min})I \tag{9.13}$$

For three-dimensional dense granular flow, $\nu_{max} = 0.6$ and $\nu_{min} = 0.5$; then the above equation is simplified as:

$$\nu = 0.6 - 0.1I \tag{9.14}$$

If I is constant, then ν is also constant.

9.8 Plane shear flow with zero gravity

Assuming that the thickness of the granular layer is h, the external stress P is acting on the top of the plane, and the speed of motion of the top layer of particles is u, we have:

$$\gamma = \frac{u}{d}, \quad I = \frac{\gamma d}{\sqrt{P/\rho_p}} = \frac{ud}{h\sqrt{P/\rho_p}} \tag{9.15}$$

Because gravity is ignored, P is evenly distributed within the granular flow, I is constant, and the velocity distribution within the granular flow is shown in Figure 9.14.

$$v(z) \propto z, \quad \phi(z) = \text{const.} \tag{9.16}$$

9.8.2 Slope flow under gravity

As shown in Figure 9.15, assuming that the plane is inclined at an angle θ and the speed of motion of surface particles is u, the following relationships can be established:

$$P = \rho_p g h, \quad \gamma = \frac{u}{d}, \quad I = \frac{\gamma d}{\sqrt{P/\rho_p}} = \frac{ud}{h\sqrt{gh}} \tag{9.17}$$

When the granular flow achieves equilibrium, the ratio of tangential stress to normal stress is $\tan(\theta)$. Therefore, I is constant throughout the particle layer, and correspondingly, v is equal everywhere (v depends on the angle θ).

$$v(z) \propto z^{3/2}, \quad v = \text{constant} \tag{9.18}$$

To investigate how constitutive relations change, with respect to the above internal parameters, we propose the stress components as follows:

$$\tau_{ij} = f_{ij}\left(\dot{\gamma}, C, D, \rho, K_n, \mu, e\right) \tag{9.19}$$

For polydispersed granular materials, the mean particle diameter $\tilde{D}$ and mean stiffness $\tilde{K}_n$ replace D and K_n, respectively, in equation (9.19). Using the dimensionless stress

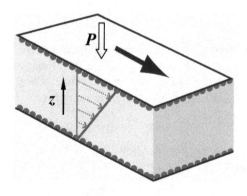

Figure 9.14: The velocity distribution of plane shear flow under zero gravity.

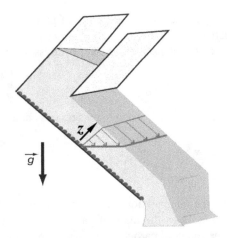

Figure 9.15: The velocity distribution of slope flow under gravity.

$\tau_{ij}^* = \tau_{ij}/\rho\tilde{D}^2\dot\gamma^2$ and dimensionless stiffness $K_n^* = \tilde{K}_n/\rho\tilde{D}^3\dot\gamma^2$, equation (9.19) can be written as:

$$\tau_{ij}^* = f_{ij}^*(K_n^*, C, \mu, e) \qquad (9.20)$$

From the observed relation between stress and shear rate, a simple power law is suggested as:

$$\tau_{ij}^* = a_{ij}\left(K_n^*\right)^{b_{ij}} \qquad (9.21)$$

where $a_{ij} = a_{ij}(C, \mu, e)$ and $b_{ij} = b_{ij}(C, \mu, e)$. As will be shown below, both a_{ij} and b_{ij} vary slowly with K_n^*. Substituting $\tau_{ij}^* = \tau_{ij}/\rho\tilde{D}^2\dot\gamma^2$ and $K_n^* = \tilde{K}_n/\rho\tilde{D}^3\dot\gamma^2$ into equation (9.21), we obtain the constitutive equation of granular materials as:

$$\tau_{ij} = a_{ij}\tilde{K}_n^{b_{ij}}(\rho\dot\gamma^2)^{1-b_{ij}}\tilde{D}^{2-3b_{ij}} \quad \text{or} \quad \tau_{ij} = a_{ij}\tilde{K}_n^{1-c_{ij}}(\rho\dot\gamma^2)^{c_{ij}}\tilde{D}^{3c_{ij}-1} \qquad (9.22)$$

When b_{ij} approaches 1, the stresses are rate independent and when b_{ij} approaches 0, the stresses agree with the kinetic theory. A rate dependency index $0 < c_{ij} = 1 - b_{ij} < 1$ is thus defined. The higher the value of c_{ij}, the more kinetic-gas-like the stress is.

Appendix A. Internal parameters of granular flows

The flow regime transition of granular materials is associated with how forces are transmitted within the assembly. Therefore, this transition should depend on the material properties at the particle level. It is desirable to determine if there exists a universal parameter(s) that controls the transition for all granular materials. In a granular flow, the contact time, coordination number, and force chain all change significantly during regime transition. Investigating these internal parameters should be the first step towards a quantitative theory for flow transition in a granular material.

With the calculation of particle contact processes of granular materials at a microscale, the contact force and fluctuation velocity of each phase can be determined to calculate the macrostresses. The macrostresses consist of the contact stress and the kinetic stress, and can be calculated as:

$$\sigma_{ij} = \sigma_{ij}^c + \sigma_{ij}^k$$

Here, σ_{ij} is the macrostress, σ_{ij}^c and σ_{ij}^k are the contact stress and the kinetic stress, and we have:

$$\sigma_{ij}^c = \frac{1}{V} \sum_{k=1}^{N} \sum_{l=1}^{N_k} \left(r_i^{lk} F_j^{kl} \right), \quad \sigma_{ij}^k = \frac{1}{V} \sum_{k=1}^{N} M_k \left(u'^k_i u'^k_j \right) \tag{A.2}$$

where V is the volume of the computational domain, N is the particle number, N_k is the contact number of particle k, M_k is the mass of particle k, r_i^{lk} is the position tensor from the center of particle l to the center of particle k, F_j^{kl} is the tensor of the total force exerted by particle l on particle k, and u'^k_i and u'^k_j are the fluctuation velocity components of particle k. To study the influence of concentrations and shear rates on phase transition, the dimensionless macrostress $\sigma^* = \sigma_{ij}/\rho D^2 \gamma^2$ and the dimensionless shear rate $B = \gamma \sqrt{\rho D^3/K_n}$ are introduced; here, γ is the shear rate, D is the particle diameter, and ρ is the particle density.

The coordination number Z is defined as the average number of contacts per particle, and can be determined as

$$n = \frac{\sum_{k=1}^{N} n_c^k}{N}. \tag{A.3}$$

Here, n is the coordination number and n_c^k is the contact number of particle k. The coordination number can express the force chain distribution spatially, and has close relationships with the macrostress and the contact time number.

In the linear viscous-elastic contact force model, the binary contact time T_{bc} of two single particles can be defined as:

$$T_{bc} = \sqrt{[M(\pi^2 + \ln^2 e)]/(4K_n)} \tag{A.4}$$

Here, e is the restitution coefficient, K_n is the normal stiffness, and M is the mean mass of two particles.

In the linear force model, e is a constant determined by the particle size and material properties, and can be used to describe the granular flow characteristics.

The contact time number is the ratio of the mean contact time to the binary contact time in the granular flow collisions, and it describes the duration of a force chain from generation to breakup,

$$m = \frac{\bar{T}_c}{T_{bc}} \tag{A.5}$$

Here, $\bar{T}_c$ is the mean contact time. The binary contact time T_{bc} is a constant based on the properties of the material; therefore, the contact time number is an effective index to denote the mean contact time. In a single collision of two particles, the contact time is the binary contact time T_{bc}, and its contact time number $m = 1$. In a multiparticle collision, the contact time will be elongated due to the action with other particles, and normally, the mean contact time number $m > 1$. If the binary contact time is subtracted from the collision time, the net contact time number m' ($= m - 1$) can be used to describe the influence of elongation in the multiple collisions.

The dimensionless Savage number N_{Sav} characterizes the relative importance of collision and friction, defined as:

$$N_{Sav} = \frac{\rho D^2 \dot{\gamma}^2}{\sigma} \tag{A.6}$$

where $\rho D^2 \dot{\gamma}^2$ is the Bagnold stress caused by binary collisions and σ is the average normal stress. In order to analyze the average shear stress and shear rate, N_{Sav} can be further extended to a generalized N_{Sav} number, i.e., $N_{Sav} = \rho D^2 \dot{\gamma}^2 / \sigma_{xy}$.

References

[1] R. A. Bagnold, 'Experiments on a gravity-free dispersion of large solid spheres in a Newtonian fluid under shear', *Proc. R. Soc. London Ser. A*, 225, 49–63 (1954).

[2] S. Ji and H. H. Shen, 'Characteristics of temporal-spatial parameters in quasi-solid-fluid phase transition of granular materials', *Chin. Sci. Bull.*, 51(6), 646–654 (2006).

[3] M. Babic, H. H. Shen and H. T. Shen, 'The stress tensor in granular shear flows of uniform, deformable disks at high solids concentrations', *J. Fluid. Mech.*, 219, 81–118 (1990).

[4] H. H. Shen and B. Sankaran, 'Internal length and time scales in a simple shear granular flow', *Phys. Rev. E*, 70, 051308 (2004).

[5] F. da Cruz, S. Emam, M. Prochnow, J. N. Roux and F. Chevoir, 'Rheophysics of dense granular materials: discrete simulation of plane shear flows', *Phys. Rev. E*, 72, 021309 (2005).

[6] G. Lois, A. Lemaitre and J. Carlson, 'Numerical tests of constitutive laws for dense granular flows', *Phys. Rev. E*, 72, 051303 (2005).

[7] C. S. Campbell, 'Granular material flows: an overview', *Powder Technol.*, 162, 208–229 (2006).

[8] C. S. Campbell, 'Stress controlled elastic granular shear flows', *J. Fluid Mech.*, 539, 273–297 (2005).

[9] C. S. Campbell, 'Granular shear flows at the elastic limit', *J. Fluid Mech.*, 465, 261–291 (2005).

[10] Y. Forterre and O. Pouliquen, 'Flows of dense granular media', *Annu. Rev. Fluid Mech.*, 40, 1–24 (2008).

[11] Y. Forterre and O. Pouliquen, 'Friction law for dense granular flows: application to the motion of a mass down a rough inclined plane', *J. Fluid Mech.*, 453, 133–151 (2002).

[12] J. F. Hazzard and K. Mair, 'The importance of the third dimension in granular shear', *Geophys. Res. Lett.*, 30(13), 41 (2003).

[13] R. M. Iverson, 'The physics of debris flows', *Rev. Geophys.*, 35, 245–296 (1997).

[14] P. Jop, Y. Forterre and O. Pouliquen, 'A constitutive law for dense granular flows', *Nature*, 441(7094), 727–730 (2006).

[15] GDR MiDi, 'On dense granular flows', *Eur. Phys. J. E*, 14, 341–365 (2004).

[16] S. Ogawa, 'Multitemperature theory of granular materials', *Proc. US Japan Seminar of Continuum Mechanical and Statistical Approaches in the Mechanics of Granular Materials*, 208, Gakujutsu Bunken Fukukai, Tokyo (1978).

[17] J. Rajchenbach, 'Some remarks on the rheology of dense granular flows: a commentary on dense granular flows by GDR MiDi', *Eur. Phys. J. E*, 14, 367–437 1(2004).

[18] J. Rajchenbach, 'Granular flows', *Adv. Phys.*, 49, 229–256 (2000).

[19] S. B. Savage, 'The mechanics of rapid granular flows', *Adv. Appl. Mech.*, 24, 289–366 (1984).

Chapter 10

Preliminary multiscale mechanics

Apart from highly excited gaseous states, it can safely be assumed that only binary interactions occur in granular systems, and theoretic considerations have been successfully applied to granular gases. For granular flows and granular solid states, they are still rather poorly understood, although great technical relevance, and innumerable empirical models have been completed.

Both exhibit obvious multiscale structure characteristics, as illustrated in Figure 6.22. Multiple elastic contacts always occur and a force chain network is formed. The mechanical behavior of a granular assembly is influenced by force networks. Strain localization is also the result of buckling of the force chain network and this creates strong discontinuities in the fluctuating velocity field. This leads to the development of shear bands and it has been shown that the inclination of such bands depends on the boundary conditions. Buckling of the strong force chains produces strong discontinuities in the fluctuating velocity field. The inclusion of information regarding force networks is a basic requisite for accurately modeling the stress–strain behavior of granular assembles.

To better understand the complex mechanical behaviors of granular materials, we should attempt to learn more details about the dynamic evolution of force networks, namely how the physical properties of force networks transport to the assembly and then couple with the assembly and how to formulate these effects? We believe that some theories should be based on experiments, and some should be based on the complete analysis of mechanical behavior, as well as by using discrete element simulations. However, the mechanical properties of force networks are far from well developed. Although study of contact forces and arrangement of particles has clearly connected the stress and strain, they are not directly related to the dynamic process from the elastic deformation to the instability destruction of a granular system. From the mechanics viewpoint, it is preferable to backtrack to the mesoscopic force network and construct the formulations of its evolution.

There are two objectives to the multiscale study of granular materials. First, this includes constructing nonlinear constitutive relations based on force network evolutions. They should be validated before engineering applications. Second, it is necessary to understand localization problems, that is, predicting both stress and strain fields for a given stress state, material properties of constituent particles, and boundary conditions. In particular, this involves understanding the criteria for shear band initiations, predicting the development of the shear bands, and predicting the characteristic width of shear bands.

10.1 Macroscopic stress and strain

As often observed in triaxial tests, during the hardening stage, granular materials behave as a continuum until the instability point where deformations begin to localize into a finite zone, known as the shear band. The material behaves as a rigid block or as a continuum until the localization point is reached where the material migrates from behaving as one block to several fragments that are split by shear bands. Studies on the stress–strain relations of granular materials start with macroscale investigations, which treat a granular material as an equivalent continuum.

Most constitutive models and theories are elastoplastic, although there are also hypoplastic models and theories, which manage to retain realism while being simpler and more explicit. Hypoplasticity represents a particular class of incrementally nonlinear constitutive models, developed specifically to predict the behavior of soils. In hypoplasticity models, unlike in elastoplasticity, the strain rate is not decomposed into elastic and plastic parts, and the models do not explicitly use the notions of the yield surface and plastic potential surface. Nevertheless, the models are capable of predicting the important features of the behavior of soils, such as the critical state, dependency of the peak strength on soil density, nonlinear behavior in the small and large strain range, dependency of the soil stiffness on the loading direction, and so on.

Both the elastoplastic theory and the hypoplastic theory are continuum mechanical models that, starting from momentum conservation, focus on the total stress σ_{ij}. Because an explicit expression appears impossible, incremental relations are constructed, expressing $\partial_t \sigma_{ij}$ in terms of σ_{ij}, velocity gradient $\nabla_j v_i$ and mass density ρ. Any such expression, either for σ_{ij} or $\partial_t \sigma_{ij}$, is referred to as a constitutive relation. Typically, neither elastoplastic nor hypoplastic theories consider energy conservation. Considerable discretion is deemed necessary to account for the variation between different materials, such as simple fluids and elastic solids. This is in contrast to the hydrodynamic theory, a powerful approach to macroscopic field theories, which was pioneered by Landau and Lifshitz. By considering energy and momentum conservation simultaneously, and by combining both with thermodynamic considerations, this approach cogently deduces the proper constitutive relation for a given energy expression. In other words, if the energy expression is known, the inclusion of energy conservation adds so many constraints for the constitutive relation that it becomes unique. The modeling discretion is hence reduced to a scalar energy expression, which is quite sufficient to account for material-specific differences, such as between simple fluids and elastic solids. Energy conservation is also considered in continuum mechanics, but the discussion does not go far enough to achieve a similar degree of cogency.

Jiang and Liu derived a hydrodynamic framework starting from transient elasticity, and called it granular solid hydrodynamics (GSH). GSH is a complete continuum mechanical theory for granular media, including explicit expressions for the energy current and the entropy production. Its underlying notion is that granular materials are elastic at rest, but become transiently elastic when the particles are agitated, such as by tapping or shearing. The GSH theory includes the true temperature as a variable, and in addition it employs a granular temperature variable to quantify the extent of agitation. A free-energy expression is provided that contains the full jamming phase diagram, in the space spanned by pressure, shear stress, density, and granular temperature. In the static limit, GSH reduces to granular elasticity, shown previously to yield realistic static stress distributions. For steady-state deformations, it is equivalent to hypoplasticity.

It has been known that the earlier theories find it difficult to describe strain localization in granular materials. Therefore, a nonlocal theory with internal length scales is needed to overcome such problems, by consideration of particle rearrangement, rolling, and translation. The micropolar and high-order gradient theories can be considered as good examples to characterize the strain localization in granular materials. The fact that internal length scales are needed requires micromechanical models or laws; however, the classical constitutive models can be enhanced through the stress invariants to incorporate the micropolar effects. Usually, this kind of model requires the introduction of a characteristic length, such as the mean particle diameter, and then the strain localization could theoretically be well observed in granular materials. Even so, in addition to the experiments and theoretical developments in soil mechanics, this kind of model needs further improvement. This kind of constitutive relation is actually the intermediate approach between the continuum theory and the micromechanical theory. For example, the micropolar model or Cosserat model mainly consider the rotation of particles, and consider the component of rotation beside the transitional motion; therefore, the strain tensor includes the rotation tensor and curvature tensor. Perhaps, the multiscale mechanics of granular materials is the ultimate method to validate continuum models as well as the micropolar model and its applicability.

10.2 Macro–micro relations

In a granular material, the micro quantities include the contact forces, the particle displacements, and the particle scale geometries. The macro quantities include the stress tensor, the strain tensor, and the fabric tensors. Efforts have been made to establish macro–micro relations linking up the physical quantities of the two scales, and to develop macro constitutive relationships in light of the microscopic behaviors. To facilitate the study of the stress–strain behavior of granular materials, various ways have been proposed to subdivide a granular assembly into microelements. For example, the Voronoi–Delaunay tessellation has been used and it proposed a mathematical description of the internal structure of a granular material consisting of particles with arbitrary convex shapes. The results are called the dual cell system (the material cell system and the space cell system), which describes the spatial approximation of granular particles. These two geometrical systems are mathematically modeled and well defined in the m-dimensional space. Li and Li (2007, 2009) proposed a contact-based tessellation to quantify the material fabric, applicable in both 2D and 3D spaces, which is convenient for static and kinematic analyses, as illustrated in Figure 10.1.

In a stress field that is uniform on the macroscale, the stress tensor can be evaluated as the average stress over the volume V as:

$$\bar{\sigma}_{ij} = \frac{1}{V}\left(\sum_{P\in V}\sum_{c\in P} v_i^{Pc} f_j^{Pc} + \sum_{P\in V} R_{ij}^P\right) \tag{10.1}$$

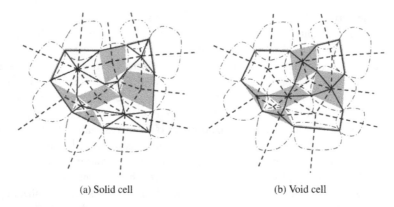

(a) Solid cell   (b) Void cell

Figure 10.1: A dual cell system in a granular assembly.

where f_j^{Pc} is the force acting at the contact point c of the particle P, and $v_i^{Pc} = X_i^P - x_i^c$ is the vector from the contact point c to the particle center X_i^P.

Equation (10.1) is referred to as the microstructural expression of the stress tensor. Note that the derivation of the stress expression is based on the Newtonian second law of motion. It is valid for both static and dynamic analyses. The expression is different from the conventional expression with the additional term $\sum_{P \in V} R_{ij}^P$ due to particle rotation.

In a strain field uniform on the macroscale, the average displacement gradient tensor over the volume V can be evaluated as:

$$\bar{e}_i = \frac{1}{V} \sum_{M \in V} \sum_{S^M} \sum_{L^{\Delta S}} x_i \Delta u_j \tag{10.2}$$

where the void cell is M, $L^{\Delta S}$ denotes the boundary of the surface increment ΔS; Δu_j denotes the variation of the normal direction of the boundary surface at point j pointing inwards, to be consistent with the sign convention in soil mechanics. In comparison to the definition of stress in equation (10.1), one of the obstacles for homogenization is that there has been no accepted general method of averaging particle displacements to obtain macroscopic strains.

The internal structure plays a crucial role in the localization process of contact forces/particle displacements from the stress/strain tensors, and in the averaging process to determine stress/strain from contact forces/particle displacements. The homogenization process is extremely complex due to the involvement of the internal structural networks, which consists of a large number of particle-scale geometrical quantities.

10.3 Multiscale mechanics

The tested stress–strain values of granular materials exhibit a high degree of scattering, for series of samples with approximately the same initial properties. It indicates that the stress–strain relations cannot be solely determined through initial averaged properties. Such uncertainty should not be simply treated as random error in experiments, but rather as the reflection of the evolution processes of internal force networks. For slight differences in force networks, during the strain process, the internal structure would evolve nonlinearly, and the difference in internal force networks is amplified.

Statistical mechanics provide a framework for relating the microscopic properties of individual atoms and molecules to the macroscopic bulk properties of materials, therefore, explaining *thermodynamics* as a result of classical and quantum-mechanical descriptions of statistics and mechanics at the microscopic level. But for granular materials, which are athermal systems, the scale couplings are in nonequilibrium, are nonlinear, and are sensitive. They cannot be treated with either statistical averaging or small perturbation, that is, they have to be treated as strong coupling. In the case of sensitive coupling, some random perturbations on small scales could be amplified, leading to intensive responses on large scales. However, current theories and approaches are still lacking in the ability to solve such problems. The statistical mesomechanics of damage may provide a paradigm for multiscale mechanics of granular matter.

The correlation between force chains and their transformations with macroscopic properties has not yet been thoroughly studied. We have introduced a mesoscale in granular matter, that is, force chains bridging the two end scales. Force chains are determined not only by the material properties of the particles at microscale, such as the Young's modulus, Poisson ratio, and the static friction coefficient, but also by macroscopic parameters, such as packing fraction, boundary conditions, and loading history. Complicated transformations of force chains would dominate macroscopic mechanical properties of granular systems. We propose a brief micro–macro framework as illustrated in Figure 6.22.

Figure 10.2 illustrates the theory framework based on the scales in Figure 6.22. From Figure 10.2, we can see that the Boltzmann equation (BE) is the basis of the kinetic theory for granular gases or very dilute granular flows. One of the most important characteristics of a real granular gas is hard binary inelastic collision. This feature of inelasticity alone is responsible for many of the qualitative differences between granular and normal gases. The transport coefficients can be obtained at the Navier–Stokes order. It is one of the most important equations of nonequilibrium statistical mechanics, namely the area of statistical mechanics that deals with systems far from thermodynamic equilibrium. Rigorously, BE is not applicable for dense granular flows in which enduring contacts among particles dominate the mechanical properties.

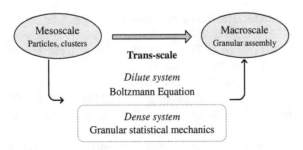

Figure 10.2: Multiscale mechanics of granular materials.

For dense granular systems, we need to investigate new statistical methods to construct a framework for coupling multitemporal-spatial scales, and explore the nonlinear, plastic, dissipation, and transition behaviors from solid-like to fluid-like. The diversity in physics and the strong coupling between multiscales and multiphysics are two of the fundamental difficulties encountered in the problem of solid failure. In addition, these difficulties can be greatly magnified by the dynamic nonlinearity and the disordered heterogeneity on multiscales. These difficulties make the popular similarity and perturbation methods unfit. Then, what is the suitable approach to this problem? We put more emphasis on the descriptions of the mesostructure of force chains, including the characteristic length, controlling parameters, and the transport to macroscale. By doing this, we could clarify how the mesodynamics is balanced with macroscale equations. These are the requirements from engineering sciences, and new methodology is expected.

In principle, the problem of deformation of granular assemblies can be represented by a statistical approach linking microscopic and macroscopic scales. However, it is difficult to represent nonequilibrium statistical evolution in a statistical approach linking mesoscopic and macroscopic scales. In addition, there are no simple direct connections between the two scales. Furthermore, a noticeable feature in the problem is the richness of structures and processes at the scale of force networks. These mesoscale structures, such as particles, play a significant role in the problem. Hence, a rational approach is to select the intermediate but essential scale, namely mesoscopic scales, and to develop a statistical approach linking mesoscopic and macroscopic scales. Such a theory can be called statistical mesoscopic mechanics of granular materials.

The mechanics of granular materials is a fascinating discipline, and granular materials widely exist in nature. This prompts us, in practical terms, to seek significant scientific explanations or solutions to problems in soils, rock fills, landslides, and in solving other practical problems, thus, contributing to the studies on granular matter mechanics. Ignoring the applications, or only considering the pure basic research of expanded knowledge, or ignoring the pure application research of expanded theoretical knowledge, is not conducive to the long-term development of granular materials mechanics. In short, we need to seek meaningful scientific problems, not only from the inherent contradictions in scientific theories but also from the actual problems.

References

[1] Q. Sun, G. Wang and K. Hu, 'Some open problems in granular matter mechanics', *Prog. Natural Sci.*, 19, 523–529 (2009).
[2] Q. Sun, F. Jin, J. Liu and G Zhang, 'Understanding force chains in dense granular materials', *Int. J. Modern Phys. B*, 24, 5743–5759 (2010).
[3] M. Xia, Y. Wei, F. Ke and Y. Bai, 'Critical sensitivity and trans-scale fluctuations in catastrophic rupture', *Pure Appl. Geophys.*, 159, 2491–2509 (2002).
[4] H. Wang, G. He, M. Xia, F. Ke and Y. Bai, 'Multiscale coupling in complex mechanical systems', *Chem. Eng. Sci.*, 59, 1677–1686 (2004).
[5] X. Zhang, X. Xu and H. Wang, 'Critical sensitivity in driven nonlinear threshold systems', *Pure Appl. Geophys.*, 161(9), 1931–1944 (2004).

WITPRESS   *...for scientists by scientists*

Aluminium Alloy Corrosion of Aircraft Structures
Modelling and Simulation

*Edited by: **J.A. DEROSE** and **T. SUTER**, EMPA (Swiss Federal Institute of Materials Science & Technology), Switzerland; **T. HACK**, EADS Innovation Works, Germany and **R.A. ADEY**, CM BEASY, UK*

Bringing together the latest research this book applies new modelling techniques to corrosion issues in aircraft structures. It describes complex numerical models and simulations from the microscale to the macroscale for corrosion of the aluminium (Al) alloys that are typically used for aircraft construction, such as AA2024. The approach is also applicable to a range of other types of structures, such as automobiles and other forms of ground vehicles.

The main motivation for developing the corrosion models and simulations was to make significant technical advancements in the fields of aircraft design (using current and new materials), surface protection systems (against corrosion and degradation) and maintenance. The corrosion models address pitting and intergranular corrosion (microscale) of Al alloys, crevice corrosion in occluded areas, such as joints (mesoscale), galvanic corrosion of aircraft structural elements (macroscale), as well as the effect of surface protection methods (anodisation, corrosion inhibitor release, clad layer, etc.).

The book describes the electrochemical basis for the models, their numerical implementation and experimental validation, and how the corrosion rate of the Al alloys at the various scales is influenced by its material properties and the surface protection methods. It will be of interest to scientists and engineers interested in corrosion modelling, aircraft corrosion, corrosion of other types of vehicle structures such as automobiles and ground vehicles, electrochemistry of corrosion, galvanic corrosion, crevice corrosion, and intergranular corrosion.

ISBN: 978-1-84564-752-0   eISBN: 978-1-84564-753-7
Published 2013 / 200pp / £148.00

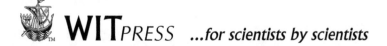
WITPRESS ...*for scientists by scientists*

Materials Characterisation VI

Computational Methods and Experiments

Edited by: **A.A. MAMMOLI**, *The University of New Mexico, USA;* **C.A. BREBBIA**, *Wessex Institute of Technology, UK and* **A. KLEMM**, *Glasgow Caledonian University, UK*

This book contains papers presented at the Sixth International Conference on the topic. Materials modelling and characterisation have become ever more closely intertwined. Characterisation, in essence, connects the abstract material model with the real-world behaviour of the material in question. Characterisation of complex materials often requires a combination of experimental and computational techniques. The conference is convened biennially to facilitate the sharing of recent work between researchers who use computational methods, those who perform experiments, and those who do both, in all areas of materials characterisation.

The papers cover such topics as: Computational Models and Experiments; Mechanical Characterisation and Testing; Micro and Macro Materials Characterisation; Corrosion Problems; Innovative Experimental Technologies; Recycled Materials; Thermal Analysis; Advances in Composites; Cementitious Materials; Structural Health Monitoring; Energy Materials.

WIT Transactions on Engineering Sciences, Vol 77

ISBN: 978-1-84564-720-9   eISBN: 978-1-84564-721-6
Forthcoming 2013 / apx 454pp / apx £195.00

CPSIA information can be obtained
at www.ICGtesting.com
Printed in the USA
LVHW061916140723
752400LV00017B/14